Environmental Risks of Non-native Conifer Monocultures in Forests

Perry

Table of contents

Chapter 6 – Do birch and rowan establish soil seed banks sufficient to compensate for a lack of seed rain after forest disturbance?

Chapter 1

General introduction

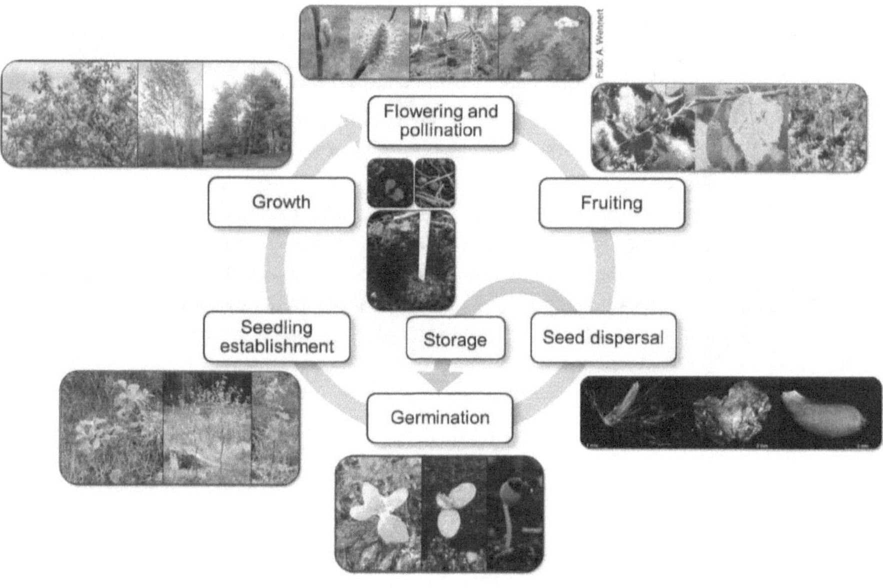

1.1 Introduction

1.1.1 Importance and relevance of the study

Since the 19th century, single-layered conifer forests, such as Norway spruce (*Picea abies* L.) and Scots pine (*Pinus sylvestris* L.) forests, have been established outside of their natural range, due to the need for fast and high-volume timber production. Once these forest sites were dominated by beech and oak (Weber & Jenssen 2006, Zerbe 2009, Löf et al. 2010). During the 20th century large open sites, the result of storms, fires, insect calamities and reparation fellings after the world wars, were reforested mainly with conifers. These forests are characterized by low species richness and homogenous structures (Profft 2013). The significant discrepancies between the tree species native and adapted to specific sites and the monocultural forests that actually inhabit them, unsuited to the conditions in which they are found, are revealed by the frequent occurrence of large-scale abiotic (snow damage, drought, storm and fire) and biotic (insect calamities) disturbances (Löf et al. 2010, Profft 2013). As climate change proceeds, the abiotic and biotic risks will intensify and the problems associated with the selection of inappropriate, conifer tree species will grow. Monocultural forests are also vulnerable to factors caused by anthropogenic activities, such as air pollution, acid rain and soil acidification. These anthropogenic threats led, for example, to the phenomenon of forest dieback in the 1980s (Krause et al. 1986, Schulze 1989). The wide range of potential disturbances are ultimately the result of the low resilience and adaptability of the current conifer stands, often leading to damage to huge areas when disturbances strike (Klimo et al. 2000, Weber & Jenssen 2006, Keidel et al. 2008, Knocke et al. 2008, Löf et al. 2010).

The consequences of rapid large-scale deforestation are the destruction of timber, the economic basis of forestry, and of habitats for numerous species. But much worse are the impacts on the complex ecosystem services forests provide for society and on the role forests play in influencing the climate. Large-scale disruption to forest ecosystems can lead to the decoupling of nutrient cycles, the modification of water regimes, the increased release of carbon dioxide and nutrients through accelerated decomposition and recomposition processes of organic matter in and on the soil, wash out, soil erosion, partial water eutrophication and the creation of opportunities for further damage by subsequent disturbances (Wohlgemuth et al. 1995, Dale et al. 2001, Seidl et al. 2014, Vilhar et al. 2014, Muscolo et al. 2017). A compensation of nutrient losses on disturbed sites may fail until the regeneration of sites occurs and a renewal of above ground litter is restored (Wohlgemuth et al. 1995, Röhrig et al. 2006). The earliest possible regeneration of disturbed areas and their ecosystem functions is vitally important, as identified in BWaldG §11 paragraph 1 and in state forest laws (e.g., SächsWaldG §20 and ThüringerWaldG §23).

Forestry has tried to counter new disturbances using proven silvicultural concepts, such as adaptation of the spatial structure of forest (e.g., series of cuts) or targeted stand development (Edelhoff 1992, Dale et al. 2001). In keeping with the goal of sustainable, multifunctional forest management, for the last 30 years a process of restoring unstable conifer forests to natural, uneven-aged, mixed broadleaf forests suited to local site conditions has been underway (Zerbe & Kreyer 2007, Ammer et al. 2008, Knocke et al. 2008, Huth et al. 2017). Knocke et al. (2008) wrote that information on the impact of the proportions of admixed tree species in coniferous forests varies between studies. It has been proven, however, that the presence of deciduous trees in coniferous forests significantly reduces the extent of the disturbances caused by storm (see Ammer et al. 2008, Clasen et al. 2008, Frischbier 2011). In mixed stands, the fact that spruce can assimilate after the deciduous trees have lost their leaves, until late autumn or winter, means the trees can develop longer crowns, which enhances their physical stability (Schütz et al. 2006, Knocke et al. 2008).

Recent extreme weather events have shown that the forestry sector must assume the inevitability of severe storms (DMG 2007, Fröhlich 2011, IPCC 2012, Kaulfuß 2012) and that these will have stark consequences in spite of the forest restoration measures that have taken place. Over the last 30 years there has been a tendency towards more frequent storm events (Rudolf & Simmer 2002, Majunke et al. 2008, Becker et al. 2016, Gregow et al. 2017), and with climate change it is to be expected that their frequency and intensity will only grow (Dale et al. 2001, Fröhlich 2011, IPCC 2012, Mölter et al. 2016). Large-scale disturbances to forests in Germany occurring since the 1990s have generated salvage wood volumes ranging from 11.0-73.7 million m³ (Majunke et al. 2008). This indicates that, in spite of all of the preventive measures undertaken by foresters, storm events will inevitably continue to cause disturbances in the future. In 2007 the storm Kyrill generated 37 million m³ of salvage wood in Germany (Majunke et al. 2008), and 30,500 ha of open forest in the federal state North Rhine-Westphalia [*Nordrhein-Westfalen*] and 11,000 ha in the federal state Thuringia [*Thüringen*]. The disturbed areas of forest were similar in size to clear-cuts (Leder et al. 2007, Thüringen Forst-AöR 2017).

Due to the increasing frequency of severe storms and their negative consequences for forest ecosystems (damage, expansion of already disturbed sites, later bark beetle attack), sustainable ecological silviculture and forest management strategies aim to achieve prompt reforestation or regeneration of disturbed sites with fast-growing tree species (Leder 2003, Brang

2005, Zerbe & Kreyer 2007, Keidel et al. 2008, Knocke et al. 2008, Löf et al. 2010, Aldinger & Kenk 2000, Unseld & Bauhus 2012, ThüringenForst 2013). For the purposes of restoration or regeneration, foresters can avail of trees already established on the site (seedling banks, trees planted under the previous canopy), soil seed banks or naturally regenerated pioneer trees. Shade-tolerant and intermediate tree species can also be planted on open sites (Röhrig et al. 2006, Ammer et al. 2008, Löf et al. 2010).

Since the observed increase in the frequency of large-scale disturbances, the integration of natural regeneration of pioneer tree species, such as *Betula* spp., *Populus tremula* L., *Salix caprea* L. and *Sorbus aucuparia* L., has been more widely incorporated in recommendations for the reforestation of open sites. The artificial reforestation (planting or seeding) of open sites is time-consuming and cost-intensive. Forestry companies rarely possess the financial and human resources necessary to ensure the clearance of all disturbed sites within a short timeframe, nor for the subsequent artificial reforestation. A cost-saving approach to reforestation is to make use of the pioneer tree species occurring naturally. A pioneer forests arises from the natural regeneration of pioneer tree species (Aldinger & Kenk 2000, Dale et al. 2001, Lässig & Motschalow 2002, Leder 2003, Brang 2005, Leder et al. 2007, Keidel et al. 2008, Löf et al. 2010, Unseld & Bauhus 2012, ThüringenForst 2013) in the course of succession, provided there are sufficient seed trees in the vicinity. Pioneer tree species are, therefore, becoming more and more important as key species in the context of forest management (Leder 2003, Leder et al. 2007, ThüringenForst 2013).

Compared to shade-tolerant climax tree species (e.g., beech and silver fir), which are less tolerant of extreme weather conditions, pioneer tree species are characterized by regular fructification, long seed dispersal distances (Perala & Alm 1990, Atkinson 1992, Zerbe 2001, Kuzovkina & Quigley 2005, Argus 2006, Hynynen et al. 2010, Żywiec et al. 2013, Fischer et al. 2016, Huth et al. 2017) and the ability to grow on open sites (Schmidt-Schütz & Huss 1998) under unfavorable weather conditions (see Renaud et al. 2010). Pioneer tree species can colonize large areas in the first year after a disturbance, thereby mitigating many of the negative consequences associated with disturbed (open) areas (Leder et al. 2007). The very fast juvenile growth of pioneer tree species compared to other species means that the sequestration of CO_2 released during and after the disturbance resumes more quickly (see Post & Kwon 2000, Aguilos et al. 2014). Climax species can naturally migrate or be planted under the protective shelter of pioneer stands (reduction of open-site climate conditions, deflection of browsing pressure, stem quality-enhancing effect for target tree species). This shelter (= forest climate) is required by climax tree species for successful establishment (Schmidt-Schütz & Huss 1998,

Aldinger & Kenk 2000, Leder et al. 2007). To create the conditions of a forest climate, these pioneer tree forests must have high stocking densities of 1,600-3,300 n ha^{-1} (Schmidt-Schütz & Huss 1998, Aldinger & Kenk 2000, Leder et al. 2007, TMLNU 2009, Hynynen et al. 2010). Forest management exploiting succession and integrating pioneer tree species facilitates a rapid regeneration of disturbed areas but can also lead to improved functioning of forest ecosystems (Knocke et al. 2008, Zerbe 2009). Pioneer tree species have a high degree of ecological importance for all forest ecosystems in terms of the diversity of species, structures and habitats (Perala & Alm 1990, Leder 1992, Raspé et al. 2000, Argus 2006, Zerbe 2009, Hynynen et al. 2010). Pioneer tree species provide habitats and food for many organisms (insects, birds, small mammals) (Kay 1985, Patterson 1993, Leder 1995, Humphrey et al. 1998, Schmidt 1998, 1999, Hacker 1999, Priha 1999, Regvar et al. 2010) and enhance soil and forest climate conditions (Horvat-Marolt 1974, Schiechtl 1992, Prien 1995, Kuzovkina & Quigley 2005, Baum et al. 2009). In this way the functional and ecological deficits (e.g., soil acidification) associated with traditional forestry (i.e., management of monocultural conifer forests) can be compensated. An increasing number and spread of pioneer tree species in forests is, therefore, both desirable and necessary (see TMLFUN 2011).

1.1.2 Research interest - regeneration ecology

The most important deciduous pioneer tree species of European temperate forests are silver and downy birch (*Betula pendula* Roth and *B. pubescens* Ehrh.), common alder (*Alnus glutinosa* (L.) Gaertn.), common aspen (*Populus tremula* L.), goat willow (*Salix caprea* L.) and rowan (*Sorbus aucuparia* L.). These species have always regenerated successfully and, as a consequence, managed to persist in Europe's managed forests even though they have not been actively cultivated.

Throughout the history of forestry there have been periods during which various target tree species have been promoted, such as spruce, pine, oaks or beech. There has never been any such promotion of deciduous pioneer tree species. Up to now in Germany, the pioneer tree species have been left to regenerate by themselves (Mantel 1990, Lang 1996, Lässig & Motschalow 2002, Röhrig et al. 2006). These species have received no attention from foresters and have always been considered secondary tree species. They occupy only very small proportions of the total forest area (Mantel 1990). In the 20th century these species were even considered weeds and removed from young stands in favor of the main tree species. Their only purpose, if indeed they had one, was to promote high quality stems amongst the target tree species during the early stages of stand development (Röhrig & Gusson 1990, Lässig & Motschalow 2002, Koski & Rousi 2005).

In the past silver fir (*Abies alba* Mill.) and common yew (*Taxus baccata* L.) were unremittingly overused by foresters but the tending and promotion measures for these species were inadequate. As a consequence, a continuous reduction of their numbers took place until they disappeared from many parts of Germany (see Pietzarka 2005, Huth et al. 2017). Today silver fir is being re-introduced on many suitable sites through seeding and planting, at considerable expense in terms of the time and financial costs involved (Kenk & Guehne 2001, Ammer et al. 2002). This example of the use and restoration of shade-tolerant tree species contrasts with that of pioneer species. Despite the difficult circumstances facing pioneer tree species over centuries, surprisingly, the species were able to survive and maintain their place in the forests, albeit with only small proportions compared to the managed target tree species (e.g., Baden Wuerttemberg [*Baden Württemberg*] 5.2 %, Hessen 7.2 %, Thuringia [*Thüringen*] 7.4 %, Bavaria [*Bayern*] 7.7 %; BMEL 2012). Clearly these light-demanding pioneer tree species have special ecological strategies that they employ within the regeneration cycle to ensure their survival under all circumstances.

However, the survival of the pioneer tree species up until today does not guarantee their continued existence in the future. The small proportions of pioneer trees still found in forests are mainly attributed to clear cutting in the past and, more recently, to disturbance events resulting in large-scale disturbed areas. Pioneer tree species are able to establish and survive on these large open sites. The question that arises, however, is how will it be possible to maintain pioneer tree species under a continuous cover forest with significantly lower numbers of large disturbed areas? To address this, it is important to understand the ecological strategies of pioneer tree species; i.e., their requirements for the successful regeneration and establishment of young trees. Only then might foresters be able to produce stand conditions suitable for the persistence of these species (= management of habitat structure for pioneer tree species).

To assess the likelihood of the natural regeneration of windthrown sites by pioneer tree species, their regeneration strategies must be understood in all aspects and for all stages of the regeneration cycle, from flowering to the established seedling (Fig. 1.1 - see Fischer et al. 2016). Empirical information about the frequency of fructification, seed quantities and the spatial and temporal distribution of seeds are central parameters in regeneration ecology. Soil seed banks and seedling banks may also be important for successful regeneration (Skoglund & Verwijst 1989, Thompson et al. 1997, Wagner & Müller-Using 1997, Raspé et al. 2000, Stancioiu & O'Hara 2006).

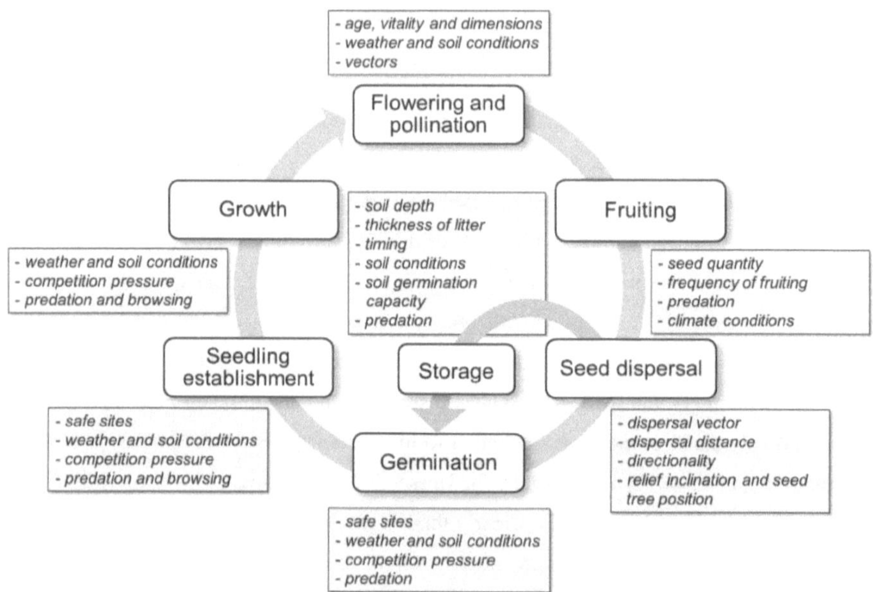

Fig. 1.1 Stages of the regeneration cycle and the main aspects for assessment of regeneration patterns and potential (based on Harper 1977).

Some information about the ecology of pioneer tree species is known from prior research, but many gaps remain and much of the information that does exist in relation to the regeneration ecology of pioneer species is contradictory. For example, it is widely held and has been shown that seed production by birch (downy and silver birch), willow (willow species) and rowan can vary between years, such as between mast and non-mast years (Sarvas 1948, 1952, Bjorkbom 1971, Houle & Payette 1990, Sperens 1997, Huth 2009, Żywiec et al. 2012), yet Cameron (1996) and Argus (2006) observed high seed production numbers annually. If pioneer tree species do not have high seed crops each year, do they employ other strategies for successful regeneration, such as by establishing a soil seed bank and seedling bank? Or are the long seed dispersal distances characteristic of pioneer tree species sufficient to compensate for low levels of seed production locally?

The ability of temperate pioneer tree species to form a soil seed bank and to regenerate from this seed reserve is discussed controversially in the literature. It has been claimed that birch seeds regenerate primarily from the annual seed rain and not from seed reserves in the soil (Hill 1979, Heinrichs 2010), yet birch seeds have also been classed as belonging to the short-term persistent soil seed bank type (viable 1-5 years) (Thompson et al. 1997). Willow seeds are short-lived and under natural conditions remain viable for only a few weeks after maturation (Lautenschlager 1994, Barsoum 2002), so that the seeds can persist in the soil for only a short period. However, this reported high rate of viability loss is contradicted by findings published by Brown & Oosterhuis (1981), Staaf et al. (1987), Bakker et al. (1996a), Bekker et al. (2000), Berger et al. (2004) and Dölle & Schmidt (2009), who recorded willow seeds in mineral soil samples. Due to embryo and seed coat dormancy, and sometimes secondary dormancy, buried rowan seeds can remain viable in soil for up to 5 years according to Erlbeck (1998). But Grime et al. 1988 (cited by Raspé et al. 2000) and Dölle & Schmidt (2009) determined that rowan seeds persist for less than a year. While rowan can build up a seedling bank under shelter (Granström 1982, Holeksa & Żywiec 2005, Żywiec & Ledwoń 2008, Heinrichs 2010), light-demanding birches and willows generally cannot (Atkinson 1992, Mihók et al. 2005). Nevertheless, goat willow and birch have also been found to grow under shelter in the understory (Skvortsov 1999, Perdereau et al. 2014, Bartsch & Rörig 2016).

The results of previous studies about the seed dispersal distribution of pioneer tree species have also revealed a high degree of variation, and occasional contradictions. Ryvarden (1971) found that about 90 % of willow seeds were deposited within 5 m of a seed tree, Gage & Cooper (2005) within 200 m, whereas Schirmer (2006) mentioned seed dispersal distances of 2-3 km. The reported maximum seed dispersal distances of birch vary between 192 m (Sarvas 1948) and 550-800 m (McEuen & Curran 2004, Huth 2009). The endozoochorous dispersal of rowan seeds by birds – the main vector for dispersal over long distances (Bakker et al. 1996b) – occurred mainly within 40 m of the seed tree (Żywiec et al. 2013). However, Leder (1992) found high rowan regeneration densities at distances of 50 m, 300-350 m and 550 m from seed trees. It should be noted here that most results pertaining to seed dispersal distances in pioneer tree species were obtained from closed forests or small gaps. Few studies have considered seed dispersal from closed forests to areas of large-scale disturbance, resulting in a corresponding lack of knowledge.

Many studies revealed the regeneration of disturbed sites by pioneer tree species to be insufficient (e.g., Lässig et al. 1995, Schmidt-Schütz & Huss 1998, Wolgemuth et al. 2002, Heurich 2009, Went 2011, Brang et al. 2015). The assessment of the natural establishment of seedlings on windthrown areas by pioneer tree species faces uncertainties due to insufficient knowledge of the different regeneration stages. Given the wide extent of the open questions identified above, it was not possible to address all of the relevant aspects of the regeneration ecology of deciduous pioneer tree species as part of this study. To fill some critical gaps in the knowledge of the regeneration cycle of pioneer tree species, an emphasis here was placed on addressing the issue of seed dispersal over large distances into disturbed sites and on the ability to build up a soil seed bank. These aspects were chosen to facilitate better decision-making processes concerning the urgency of reforestation measures (planting, seeding) and the natural regeneration of disturbed sites as part of forest management.

For the purposes of the study, the pioneer tree species occurring most frequently in Germany were chosen, namely goat willow (*Salix caprea* L.), silver birch (*Betula pendula* Roth) and rowan (*Sorbus aucuparia* L.). Between 2014 and 2017, seed dispersal studies took place on windthrown sites (storm Kyrill in 2007) on the slopes and mountain tops of the Thuringian Forest (Appendix 1, p. ii). The soil seed bank studies were located in the Tharandter Forest and Thuringian Forest (Appendix 1 and 2, p. ii). Information about the distribution and numbers of pioneer seed trees in the stands around the windthrown sites was also deemed to be of high importance (Appendix 3, p. iii). The results of the study should contribute to better regeneration scenarios for disturbed sites (success or failure) by pioneer tree species, especially following storms. Adopting a pro-active approach, the regeneration scenarios should allow for the introduction of pioneer trees to stands as an advance mitigation measure before disturbances occur.

1.3 Aims, scope and hypotheses

The investigations of the regeneration ecology of pioneer tree species undertaken as part of this study focused on (i) spatial and temporal seed distribution patterns of *Salix caprea*, *Betula pendula* and *Sorbus aucuparia* on windthrown sites, and (ii) seed storability of *Salix caprea*, *Betula pendula* and *Sorbus aucuparia* in the soil of closed coniferous forests. In order to determine the link between deposited seed densities on the ground and seed dispersal distances, it was necessary to consider the influence of the number of seed trees, their spatial distribution and the relief of the sites (valley, slope, plateau). To analyze seed densities in soil seed banks, the storage duration of seeds in the soil was tested and the dependence of seed densities in the soil on seed tree numbers studied. To limit the broad scope of the work, investigations of established young trees on disturbed sites (= seedling banks) was not part of the research.

The overarching aim was to obtain knowledge of the relevant aspects of regeneration ecology. The findings should allow for the determination of a theoretical minimum seed tree density required in forests to ensure silvicultural sufficient seed input and subsequent successful regeneration in the event of a disturbance. This requires information about the spatial distribution patterns of seeds around seed trees. From the results general silvicultural recommendations shall be derived for the inclusion (conservation and promotion) of seed trees of pioneer species in managed forests. Where there is no integration of pioneer tree species in managed stands, or this integration fails, the sites of large-scale disturbances will in future have to be artificially reforested (with high resource inputs).

There were four hypotheses guiding the study objectives and the research questions:

1) *Due to varying annual weather conditions, pioneer tree species do not produce high seed numbers every year (Sarvas 1948, Kelly & Sork 2002, Gage & Cooper 2005, Żywiec et al. 2012). It is assumed that in some years seed production may be too low to ensure successful regeneration.*

> The fructification process of pioneer tree species is influenced by site and climate conditions, individual vitality and tree dimensions. These factors influence the frequency of and the intervals between good seed crops, as well as the amount of seed produced (Kullmann 1993, Sperens 1997, Kelly & Sork 2002, Żywiec et al. 2012). Unfavorable weather conditions have negative effects on fructification, similar to the effect on heavy-seeded tree species (Wohlgemuth et al. 2016).

2) *Seeds of pioneer tree species are not omnipresent on all sites, independent of the presence of seed trees. The seed numbers deposited on disturbed sites vary considerably due to limited seed dispersal (= distance to seed tree) and environmental conditions (= relief, wind and vegetation cover) (Hill & Stevens 1981, Leder 1992, Stoyan & Wagner 2001).*

2a) *The numbers of seeds of pioneer tree species deposited on windthrown sites depend on the distances to seed trees. The species-specific seed numbers decrease rapidly with increasing distance (Leder 1992, Huth 2009). It is assumed that sufficient and dense regeneration cannot be expected on open sites located far from seed sources. Considering the high seedling mortality rates, the seed numbers necessary for successful regeneration (see Sarvas 1948) are found only in the vicinity of* Salix caprea, Betula pendula *and* Sorbus aucuparia *seed trees (Ryvarden 1971, Huth 2009, Żywiec et al. 2013).*

> Although seeds of pioneer tree species are dispersed over large distances, all anemochorously dispersed seeds (birches and willows) have a species-specific sinking rate (see Kohlermann 1950). Endozoochorously distributed seeds (rowan) are subject to the action range of the dispersal vector (e.g., small mammals and frugivorous birds) (Bakker et al. 1996b, Kollmann 2000). Therefore, pioneer tree species have long but not unlimited dispersal distances (Huth 2009, Żywiec et al. 2013).

2b) *The spatial patterns of anemochorously distributed seeds are influenced by wind (direction, speed and turbulence) and the relief (Hill & Stevens 1981, Daniels 2001, Wagner et al. 2004, Moon et al. 2013). It is assumed that seeds are dispersed over longer distances in the main wind direction than against the main direction. Therefore, higher seed densities must be found at greater distances from seed trees in the main wind direction than in the opposite direction (Stoyan & Wagner 2001, Wagner et al. 2004, Wright et al. 2008). Relief-induced downhill seed dispersal should also lead to longer seed dispersal distances (Hill & Stevens 1981).*

> The very light and small seeds are non-randomly dispersed by wind (Greene & Johnson 1996), with the result that seed densities are not equal in all azimuth directions around a seed tree (= isotropy). The seed shadow should reflect the main wind direction, which is shown by directionality (= anisotropy). However, seed dispersal on disturbed sites is also influenced by lateral winds and wind turbulence, normal features of open sites (Kohlermann 1950, Moon et al. 2013). Turbulence may affect the main wind direction and the directionality of seed dispersal observed in closed forests (Stoyan & Wagner 2001, Wagner et al. 2004, Wright et al. 2008, Huth 2009).
>
> It is assumed that the relief-induced lengthening (downhill) or shortening (uphill) of seed dispersal distances are caused by slope inclination in combination with the sinking rate of the seeds.

3) *Seeds of* Betula pendula *and* Sorbus aucuparia *are storable in the soil, enabling these species to build up a soil seed bank (Granström & Fries 1985, Erlbeck 1998). Therefore, these pioneer tree species should be able to regenerate from their seed reserves in the soil. The short-lived goat willow seeds (Lautenschlager 1994), on the other hand, are not capable of establishing a soil seed bank.*

> Temperate tree species lack the ability to form a long-term persistent soil seed bank (Donelan & Thompson 1980, Bossuyt & Hermy 2001), but birch and rowan should be able to form a short-term persistent soil seed bank (Thompson et al. 1997). This seed reserve in the soil could facilitate the regeneration of disturbed sites even if there is no birch or rowan seed rain (i.e., pioneer seed trees were also felled by storm).

1.4 Study outline

The study entitled '*The ability of pioneer tree species to mitigate the effects of site disturbance by fast and effective natural regeneration*' comprises seven chapters, including this general introduction to the study and the topic (chapter 1).

In the following chapters, the most important aspects of the regeneration ecology studied in relation to the pioneer tree species silver birch, goat willow and rowan are presented. However, as it was not possible to present all of the results obtained over the course of the study in the research papers making up chapters 2-6, in the following short chapter summaries the major aspects excluded from the papers and the corresponding results are also briefly presented (in italics).

In chapter 2 the results of the study of *Salix caprea* seed dispersal from spruce forests onto five windthrown forest sites in the Thuringian Forest, Germany, are presented. The study took place in 2015 and 2016 using sticky, non-drying seed traps. The influence of seed tree number, relief and directionality on seed shadow were considered in this study. A parentage analysis of the saplings established on a windthrown site was performed to substantiate the seed trap data.

In chapter 3 the results of uphill and downhill *Betula pendula* seed dispersal on two windthrown forest sites in the Thuringian Forest, Germany, are presented. The study took place in 2015 and 2016 using funnel shaped seed traps. Inverse modelling of isotropic and anisotropic seed dispersal were used to analyze the influence of seed crop, relief, seed tree numbers and the position of the seed trees (valley, slope or plateau).

Chapter 4 contains the analysis from 2015 of the impact of structural elements at five windthrown forest sites in the Thuringian Forest, Germany, on endozoochorous seed dispersal by birds. Combined knowledge of the behavior of frugivorous bird species and plant characteristics is necessary to predict the effectiveness of endozoochorous seed dispersal; for example, of rowan.

Data on Sorbus aucuparia *seed dispersal by birds could not be presented as it was originally intended. The fructification of rowan seed trees in the Thuringian Forest was very low in the study years 2014 to 2016. As a consequence, rowan seeds were not found in the bird droppings collected on the dropping traps on all study sites. Only the spatial patterns of bird droppings on windthrown forest sites were analyzed.*

Chapter 5 contains a review of soil seed banks of deciduous pioneer tree species (*Betula* spp., *Alnus glutinosa*, *Salix* spp., *Populus tremula* and *Sorbus aucuparia*) in European temperate forests. The review summarizes general findings on the ability of pioneer tree species to form a soil seed bank and highlights gaps in the knowledge.

Chapter 6 is based on the conclusions of the soil seed bank review presented in chapter 5. The chapter contains the details of an artificial seed burial experiment for seeds of *Betula pendula* and *Sorbus aucuparia* established in a conifer forest in the Tharandter Forest, Germany, and a soil seed bank investigation for silver birch in spruce forests in the Thuringian Forest and Tharandter Forest, Germany. The burial experiment focused on the seed storability of silver birch and rowan in soil, and their ability to regenerate from a soil seed bank in a temperate forest. The soil seed bank investigation looked at seed densities in the soil with regard to a variation in the number of seed sources.

The results obtained for the artificially buried seeds of Salix caprea *were not presented in the chapter, because no seeds remained viable after the first storage period of six months. The results of the soil seed bank investigation for rowan seeds in the soil around individual seed trees in a spruce forest were also omitted due to a lack of viable rowan seeds in the soil.*

Finally, chapter 7 features a general discussion of all of the findings of the study and the derived recommendations for silvicultural practice to ensure the conservation and revitalization of seed trees of pioneer tree species and to promote the regeneration of disturbed forest sites using pioneer tree species.

1.5 References

Aguilos M, Takagi K, Liang N, Ueyama M, Fukuzawa K, Nomura M, Kishida O, Fukazawa T, Takahashi H, Kotsuka C, Sakai R, Ito K, Watanabe Y, Fujinuma Y, Takahashi Y, Murayama T, Saigusa N, Sasa K (2014): Dynamics of ecosystem carbon balance recovering from a clear-cutting in a cool-temperate forest. Agricultural and Forest Meteorology 197: 26-39.

Aldinger E, Kenk G (2000): Natürliche Wiederbewaldung von Sturmflächen. Merkblätter der Forstlichen Versuchs- und Forschungsanstalt Baden-Württemberg 51: 9 S.

Ammer C, Mosandl R, El Kateb H (2002): Direct seeding of beech (*Fagus sylvatica* L.) in Norway spruce (*Picea abies* [L.] Karst) stands - effects of canopy density and fine root biomass on seed germination. Forest Ecology of Management 159: 59-72.

Ammer C, Bickel E, Kölling C, (2008): Converting Norway spruce stands with beech - a review of arguments and techniques. Austrian Journal of Forest Science 125(1): 3-26.

Argus GW (2006): Guide to *Salix* (willow) in the Canadian Maritime Provinces (New Brunswick, Nova Scotia, and Prince Edward Island). E-Publishing, Ontario. URL:

http://accs.uaa.alaska.edu/files/botany/publications/2006/GuideSalixCanadianAtlanticMaritime.pdf. Accessed 10 February 2018.

Atkinson MD (1992): *Betula pendula* Roth (*B. verrucosa* Ehrh.) and *B. pubescens* Ehrh. Journal of Ecology 80(4): 837-870.

Baum C, Leinweber P, Weih M, Lamersdorf N, Dimitriou I (2009): Effects of short rotation coppice with willows and poplar on soil ecology. Agriculture and Forestry Research 59: 183-196.

Bakker JP, Bakker ES, Rosén E, Verwej GL, Bekker RM (1996a): Soil seed bank composition along a gradient from dry alvar grassland to Juniperus shrubland. Journal of Vegetation Science 7: 165-176.

Bakker JP, Poschlod P, Strykstra RJ, Bekker RM, Thompson K (1996b): Seed banks and seed dispersal: important topics in restoration ecology. Acta Botanica Neerlandica 45(4): 461-490.

Barsoum N (2002): Relative contributions of sexual and asexual regeneration strategies in *Populus nigra* and *Salix alba* during the first years of establishment on a braided gravel bed river. Evolutionary Ecology 15: 255-279.

Bartsch N, Röhrig E (2016): Waldökologie - Einführung für Mitteleuropa. Springer-Verlag, Berlin, Heidelberg, 417 S.

Becker A, Hafer M, Junghänel T, Müller HJ, Sterker C, Walawender E, Weigl E, Winterrath T (2016): Bewertung des Starkregenrisikos in Deutschland auf der Basis von Radardaten. 10. DWD Klimatagung, 3. November 2016 in Offenbach am Main

Bekker RM, Verwej GL, Bakker JP, Fresco LFM (2000): Soil seed bank dynamics in hayfield succession. Journal of Ecology 88: 594-607.

Berger TW, Sun B, Glatzel G (2004): Soil seed banks of pure spruce (*Picea abies*) and adjacent mixed species stands. Plant and Soil 264: 53-67.

BMEL (2012): Dritte Bundeswaldinventur. URL: https://bwi.info/start.aspx. Abruf: 05.05.2019.

Bjorkbom JC (1971): Production and germination of paper birch seed and its dispersal into a forest opening. Northeastern Forest Experiment Station, Forest Service Research Paper NE-209, U.S. Department of Agriculture, Upper Darby, USA 15 p.

Bossuyt B, Hermy M (2001): Influence of land use history on seed banks in European temperate forest ecosystems: a review. Ecography 24(2): 225-238.

Brang P (2005): Verteilung der Naturverjüngung auf großen Lothar-Sturmflächen. Schweizerische Zeitschrift für Forstwesen 156: 467-476.

Brang P, Hilfiker S, Wasem U, Schwyzer A, Wolgemuth T (2015): Langzeitforschung auf Sturmflächen zeigt Potenzial und Grenzen der Naturverjüngung. Schweizerische Zeitschrift für Forstwesen 166(3): 147-158.

Brown AHF, Oosterhuis L (1981): The role of buried seed in coppicewoods. Biological Conservation 21: 19-38.

Cameron AD (1996): Managing birch woodlands for the production of quality timber. Forestry 69(4): 357-371.

Clasen C, Frischbier N, Zehnert T (2008): Ursachenanalyse zum Schadausmaß des Sturmes „Kyrill" in Thüringen. AFZ-Der Wald (14): 746-748.

Dale VH, Joyce LA, McNulty S, Neilson RP, Ayres MP, Flannigan MD, Hanson PJ, Irland LC, Lugo AE, Peterson CJ, Simberloff D, Swanson FJ, Stocks BJ, Wotton BM (2001): Climate change and forest disturbances. BioScience 51(9): 723-734.

Daniels J (2001): Ausbreitung der Moorbirke (*Betula pubescens* Ehrh. agg.) in gestörten Hochmooren der Diepholzer Moorniederung. Osnabrücker Naturwissenschaftliche Mitteilungen 27: 39-49.

DMG (2007): Stellungnahme der Deutschen Meteorologischen Gesellschaft zur Klimaproblematik. Mitteilungen der Deutschen Meteorologischen Gesellschaft Heft 4: 12-14.

Dölle M, Schmidt W (2009): The relationship between soil seed bank, above-ground vegetation and disturbance intensity on old-field successional permanent plots. Applied Vegetation Science 12: 415-428.

Donelan M, Thompson K (1980): Distribution of buried viable seeds along a successional series. Biological Conservation 17: 297-311.

Edelhoff A (1992): Forstlicher Rückblick auf die Sturmkatastrophe 1990 - Daten, Analysen, waldbauliche Konsequenz. In: LAFO (Hrsg.): Sturmschäden 1990. Schriftenreihe der Landesanstalt für Forstwissenschaft Nordrhein-Westfalen, Bd. 5: 59-75.

Erlbeck R (1998): Die Vogelbeere (*Sorbus aucuparia*) - ein Porträt des Baumes des Jahres 1997. In: Schmidt O (Hrsg.): Beiträge zur Vogelbeere. Berichte aus der Bayrischen Landesanstalt für Wald und Forstwirtschaft (LWF), Nr. 17: 2-14.

Falińska K (1999): Seed bank dynamics in abandoned meadows during a 20-year period in the Białowieża National Park. Journal of Ecology 87(3): 461-475.

Fischer H, Huth F, Hagemann U, Wagner S (2016): Developing restoration strategies for temperate forests using natural regeneration processes. In: Stanturf JA (eds): Restoration of boreal and temperate forests. CRC Press, Boca Raton, 2nd edition, pp. 103-164.

Frischbier N (2011): Klimawandel in Wald und Forstwirtschaft. In: TLUG (Hrsg.): Naturschutz im Wandel, Nr. 94: 40-51.

Fröhlich D (2011): Stürmische Gesellen: Lothar, Kyrill & Co.- Zur Problematik, künftige Entwicklung von Winterstürmen abzuschätzen. LWF aktuell 80: 38-40.

Gage EA, Cooper DJ (2005): Patterns of willow seed dispersal, seed entrapment, and seedling establishment in a heavily browsed montane riparian ecosystem. Canadian Journal of Botany 83: 678-687.

Granström A (1982): Seed bank in five boreal forest stands originating between 1810 and 1963. Canadian Journal of Botany 60(9): 1815-1821.

Granström A (1987): Seed viability of fourteen species during five years of storage in a forest soil. Journal of Ecology 75(2): 321-331.

Granström A, Fries C (1985): Depletion of viable seeds of *Betula pubescens* and *Betula verrucosa* sown onto some north Swedish forest soils. Canadian Journal of Forest Research 15(6): 1176-1180.

Greene DF, Johnson EA (1996): Wind dispersal of seeds from a forest into a clearing. Ecology 77: 595-609.

Gregow H, Laaksonen A, Alper ME (2017): Increasing large scale windstorm damage in Western, Central and Northern European forests, 1951-2010. Scientific Reports 7/ 46397: 1-7.

Hacker H (1999): Die Insektenwelt der Weiden. In: LWF (Hrsg.): Beiträge zur Silberweide. Berichte aus der Bayrischen Landesanstalt für Wald und Forstwirtschaft (LWF), Tagungsband zum Baum des Jahres 1999: 25-27.

Harper JL (1977): Population biology of plants. Academic Press, New York, 892 p.

Heinrichs S (2010): Response of the understorey vegetation to select cutting and clear cutting in the initial phase of Norway spruce conservation. Georg-August-Universität Göttingen, Mathematisch-Naturwissenschaftlichen Fakultäten, book, 161 p.

Heurich M (2009): Progress of forest regeneration after a large-scale *Ips typographus* outbreak in the subalpine *Picea abies* forests of the Bavarian Forest National Park. Silva Gabreta 15(1): 49-66.

Hill MO (1979): The development of a flora in even-aged plantations. In: Ford ED, Malcolm DC, Atterson J (Eds): The ecology of even-aged forest plantations. Institute of Terrestrial Ecology, Cambridge, pp. 175-192.

Hill MO, Stevens PA (1981): The density of viable seed in soils of forest plantations in upland Britain. Journal of Ecology 69: 693-709.

Holeksa J, Żywiec M (2005): Spatial pattern of a pioneer tree seedling bank in old-growth European subalpine spruce forest. Ekológia (Bratislava) 24(3): 263-276.

Holm SO (1994). Reproductive patterns of *Betula pendula* and *B. pubescens* coll. along a regional altitudinal gradient in northern Sweden. Ecography 17(1): 60-72.

Horvat-Marolt S (1974): Pionirski gozd in iva kot Pionirska drevesna Vrsta II. del - Konkurenčna moč ive (*S. caprea*) kot pionirja v pedosferi (Pioneer forest and sallow (*Salix caprea*) as a pioneer tree species II. Competitive strength of the sallow as a pioneer in the pedosphere). Zb Gozdarstva Lesar. 12: 5-40.

Houle G, Payette S (1990): Seed Dynamics of *Betula alleghaniensis* in a deciduous forest of north- eastern North America. Journal of Ecology 78: 677-690.

Houle G (1998): Seed dispersal and seedling recruitment of *Betula alleghaniensis*: Spatial inconsistency in time. Ecology 79: 807-818.

Humphrey JW, Holl K, Broome AC (1998): Birch in spruce plantations - Management for Biodiversity. Forestry Commission Technical Paper 26, 78 p.

Huth F (2009): Untersuchungen zur Verjüngungsökologie der Sand-Birke (*Betula pendula* Roth). Technischen Universität Dresden, Fakultät Forst-, Geo- und

Hydrowissenschaften, Tharandt, book, 383 S.

Huth F, Wehnert A, Tiebel K, Wagner S (2017): Direct seeding of Silver fir (*Abies alba* Mill.) to convert Norway spruce (*Picea abies* L.) forests in Europe: A review. Forest Ecology and Management 403: 61-78.

Hynynen J, Niemistö P, Viherä-Aarnio A, Brunner A, Hein S, Velling P (2010): Silviculture of birch (*Betula pendula* Roth and *Betula pubescense* Ehrh.) in northern Europe. Forestry 83(1): 103-119.

IPCC (2012): Managing the risks of extreme events and disasters to advance climate change adaptation. A special report of working groups I and II of the intergovernmental panel on climate change [Field CB, Barros V, Stocker TF, Qin D, Dokken DJ, Ebi KL, Mastrandrea MD, Mach KJ, Plattner GK, Allen SK, Tignor M, Midgley PM (Eds)]. Cambridge University Press, Cambridge, UK and New York, 582 p.

Karlsson M (2001): Natural regeneration of broadleaved tree species in southern Sweden - Effects of silvicultural treatments and seed dispersal from surrounding stands. Swedish University of Agricultural Sciences, Southern Swedish Forest Research Centre, Alnarp, PhD book, 44 p.

Kaulfuß S (2012): Nach dem Sturm ist vor dem Sturm - Wie senke ich das Sturmrisiko meines Waldes. Forstliche Versuchs- und Forschungsanstalt Baden-Württemberg (FVA), Freiburg.

Kay QON (1985): Nectar from willow catkins as a food source for Blue Tits. Bird Study 32: 40-44.

Keidel S, Meyer P, Bartsch N (2008): Untersuchungen zur Regeneration eines naturnahen Fichtenwaldökosystems im Harz nach großflächiger Störung. Forstarchiv 79: 187-196.

Kelly D, Sork VL (2002): Mast seeding in perennial plants: why, how, where? Annual Review of Ecology, Evolution, and Systematics 33: 427-47.

Kenk G, Guehne S (2001): Management of transformation in central Europe. Forest Ecology and Management 151: 107-119.

Klimo E, Hager H, Kulhavý J (Eds) (2000): Spruce monocultures in Central Europe - Problems and prospects. EFI Proceedings No. 33, 208 p.

Knocke T, Ammer C, Stimm B, Mosandl R (2008): Admixing broadleaved to coniferous tree species: a review on yield, ecological stability and economics. European Journal of Forest Research 127: 89-101.

Kohlermann L (1950): Untersuchungen über die Windverbreitung der Früchte und Samen mitteleuropäischer Waldbäume. Forstwissenschaftliches Centralblatt 69: 606-624.

Kollmann J (2000): Dispersal of fleshy-fruited species: a matter of spatial scale? Perspectives in Plant Ecology, Evolution and Systematics 3: 29-51.

Komulainen M, Vieno M, Yarmishko VT, Daletskaja TD, Maznaja EA (1994): Seedling establishment from seeds and seed banks in forests under long-term pollution stress: a potential for vegetation recovery. Canadian Journal of Botany 72(2): 143-149.

Koski V, Rousi M (2005): A review of the promises and constraints of breeding Silver birch (*Betula pendula* Roth) in Finland. Forestry 78(2): 187-198.

Krause GHM, Arndt U, Brandt CJ, Bucher J, Kenk G, Matzner E (1986): Forest decline in Europe: development and possible causes. Water, Air, and Soil Pollution 31(3-4): 647-668.

Kullmann L (1993): Tree limit dynamics of *Betula pubescens* ssp. *tortuosa* in relation to climate variability: evidence from central Sweden. Journal of Vegetation Science 4: 765-772.

Kuzovkina YA, Quigley MF (2005): Willows beyond wetlands: uses of *Salix* L. species for environmental projects. Water, Air, & Soil Pollution 162: 183-204.

Lang P (1996): Streifzug durch die Forstgeschichte. In: Hatzfeld H Graf (Hrsg.): Ökologische Waldwirtschaft - Grundlagen-Aspekte-Beispiele. C.F. Müller Verlag, Heidelberg, S. 15-36.

Lässig R, Egli S, Odermatt O, Schönenberger W, Stöckli B, Wohlgemuth T (1995): Beginn der Wiederbewaldung auf Windwurfflächen. In: Schweizerische Forstverein (Hrsg.): Entwicklung von Windwurfflächen in der Schweiz - Erste Forschungsergebnisse, Beobachtungen und Erfahrungen. Schweizerische Zeitschrift für Forstwesen 146: 893-911.

Lässig R, Motschalow SA (2002): Wiederbewaldung von belassenen und geräumten Sturmwurfflächen im Mittel-Ural. In: Hessisches Ministerium für Umwelt, Landwirtschaft und Forsten (Hrsg.): Natürliche Entwicklung von Wäldern nach Sturmwurf - 10 Jahre Forschung im Naturwaldreservat Weiherskopf, Naturwaldreservate in Hessen 38: 23-34.

Lautenschlager D (1994): Die Weiden von Mittel- und Nordeuropa. Birkhäuser Verlag, Basel. überarbeitete Auflage. 171 S.

Leder B (1992): Weichlaubhölzer- Verjüngungsökologie, Jugendwachstum und Bedeutung in Jungbeständen der Hauptbaumarten Buchen und Eiche. Schriftenreihe der Landesanstalt für Forstwissenschaft Nordrhein-Westfalen, Sonderband, 416 S.

Leder B (1995): Jugendwachstum und waldbauliche Behandlung von natürlichen angesamten Weichlaubhölzern in Laubholzjungwüchsen. In: LÖBF (Hrsg.): Weichlaubhölzer und Sukzessionsdynamik in der naturnahen Waldwirtschaft. Schriftenreihe der Landesanstalt für Ökologie, Bodenordnung und Forst/Landesamt für Agrarordnung Nordrhein-Westfalen 4: 29-44.

Leder B (2003): „Natürliche Wiederbewaldung nach Fichtenwindwurf 1990". LÖBF-Mitteilungen 2: 40-43.

Leder B, Asche N, Dame G, Gertz M, Hein F, Kreienmeier U, Naendrup G, Sondermann P, Spelsberg G, Stemmer M, Wagner H-C, Wrede Ev (2007): Empfehlung für die Wiederbewaldung der Orkanflächen in Nordrhein-Westfalen, Arnsberg. Landesbetrieb Wald und Holz Nordrhein-Westfalen, Lehr- und Versuchsforstamt Arnsberger Wald (Hrsg), 71 S.

Löf M, Bergquist J, Brunet J, Karlsson M, Welander NT (2010): Conversion of Norway spruce stands to broadleaved woodland – regeneration systems, fencing and performance of planted seedlings. Ecological bulletins 53: 165-173.

Majunke C, Matz S, Müller M (2008): Sturmschäden in Deutschlands Wäldern von 1920 bis 2007. AFZ-Der Wald 7: 380-381.

Mantel K (1990): Wald und Forst in der Geschichte: ein Lehr- und Handbuch. Verlag M. & H. Schaper, Hannover, 518 S.

McEuen AB, Curran LM (2004): Seed dispersal and recruitment limitation across spatial scales in temperate forest fragments. Ecology 85(2): 507-518.

Mihók B, Gálhidy L, Kelemen K, Standovár T (2005): Study of gap-phase regeneration in a managed beech forest: relations between tree regeneration and light, substrate features and cover of ground vegetation. Acta Silvatica & Lignaria Hungarica 1: 25-38.

Mölter T, Schindler D, Albrecht AT, Kohnle U (2016): Review on the projections of future storminess over the north Atlantic European region. Atmosphere 7(60): 1-40.

Moon K, Duff TJ, Tolhurst KG (2013): Characterising forest wind profiles for utilisation in fire spread models. In: Piantadosi J, Anderssen RS, Boland J (Eds): MODSIM2013, Modelling and Simulation Society of Australia and New Zealand. Presented at the 20th International Congress on Modelling and Simulation, Adelaide, Australia, pp. 214-220.

Muscolo A, Settineri G, Bagnato S, Mercurio R, Sidari M (2017): Use of canopy gap openings to restore coniferous stands in Mediterranean environment. iForest 10: 322-327.

Osumi K, Sakurai S (2002): The unstable fate of seedlings of the small-seeded pioneer tree species, *Betula maximowicziana*. Forest Ecology and Management 160: 85-95.

Patterson GS (1993): The value of birch in upland forests for wildlife conservation. Forestry Commission Bulletin 109, 34 p.

Perdereau AC, Kelleher CT, Douglas GC, Hodkinson TR (2014): High levels of gene flow and genetic diversity in Irish populations of *Salix caprea* L. inferred from chloroplast and nuclear SSR markers. Plant Biology 14: 1-12.

Perala DA, Alm AA (1990): Reproductive ecology of birch: A review. Forest Ecology and Management 32(1): 1-38.

Pietzarka U (2005): Zur ökologischen Strategie der Eibe (*Taxus baccata* L.) - Wachstums- und Verjüngungsdynamik. Forstwissenschaftliche Beiträge Tharandt 25, Eugen Ulmer GmbH & Co., Stuttgart, 11 S.

Post WM, Kwon KC (2000): Soil carbon sequestration and land-use change: processes and potential. Global Change Biology 6: 317-328.

Prien S (1995): Struktur, waldbauliche Eigenschaften und waldbauliche Behandlung von Ebereschenvorwäldern. In: LÖBF (Hrsg.): Weichlaubhölzer und Sukzessionsdynamik in der naturnahen Waldwirtschaft. Schriftenreihe der Landesanstalt für Ökologie,

Bodenordnung und Forst/ Landesamt für Agrarordnung Nordrhein-Westfalen, Bd. 4: 45-59.

Priha O (1999): Microbial activities in soils under Scots pine, Norway spruce and Silver birch. Finnish Forest Research Institute, Research Papers No. 731, 51 p.

Profft I (2013): Waldumbau in den mittleren, Hoch- und Kammlagen des Thüringer Waldes. Auftaktkolloquium - Waldumbau in den mittleren, Hoch- und Kammlagen des Thüringer Waldes. ThüringenForst-Anstalt öffentlichen Rechts, Gotha, 28.02.2013.

Raspé O, Findlay C, Jacquemart A-L (2000): *Sorbus aucuparia* L. Journal of Ecology 88: 910-930.

Regvar M, Likar M, Piltaver A, Kugonic N, Smith JE (2010): Fungal community structure under goat willows (*Salix caprea* L.) growing at metal polluted site: the potential of screening in a model phytostabilisation study. Plant Soil 330: 345-356.

Renaud V, Innes JL, Dobbertin M, Rebetez M (2010): Comparison between open-site and below-canopy climatic conditions in Switzerland for different types of forests over 10 years (1998-2007). Theoretical and Applied Climatology 105(1-2): 119-127.

Röhrig E, Gussone HA (1990): Waldbau auf ökologischer Grundlage. Band 2: Baumartenwahl, Bestandesbegründung und Bestandespflege. Verlag Paul Parey, Hamburg - Berlin, 6th ed., 314 S.

Röhrig E, Bartsch N, Lüpke Bv (2006): Waldbau auf ökologischer Grundlage. Eugen Ulmer KG, Stuttgart (Hohenheim). 7. Aufl., 479 S.

Rudolf B, Simmer C (2002): Niederschlag, Starkregen und Hochwasser. Deutscher Wetterdienst (DWD), 16 S.

Ryvarden L (1971): Studies in seed dispersal. I. Trapping of Diaspores in the Alpine Zone at Finse, Norway. Norwegen Journal of Botany 18: 215-226.

Sarvas R (1948): A research on the regeneration of birch in South Finland. Communicationes Instituti Forestalis Fenniae 35(4): 82-91.

Sarvas R (1952): On the flowering of birch and the quality of seed crop. Communicationes Instituti Forestalis Fenniae 40: 1-35.

Schiechtl HM (1992): Weiden in der Praxis - Die Weiden Mitteleuropas, ihre Verwendung und ihre Bestimmung. Patzer-Verlag, Berlin-Hannover, 130 S.

Schirmer R (2006): *Salix alba* Linné. In: Schütt P, Weisgerber H, Schuck HJ, Lang U, Stimm B. Roloff A (Eds): Enzyklopädie der Laubbäume. Nikol Verlagsgesellschaft mbH & Co KG, Hamburg, 535-550.

Schmidt O (1998): Vogelbeere und Tierwelt. In: Schmidt O (Hrsg.): Beiträge zur Vogelbeere. Berichte aus der Bayrischen Landesanstalt für Wald und Forstwirtschaft (LWF), Nr. 17: 59-64.

Schmidt O (1999): Vogelwelt und Weiden. In: LWF (Hrsg.): Beiträge zur Silberweide. Berichte aus der Bayrischen Landesanstalt für Wald und Forstwirtschaft (LWF), Tagungsband zum Baum des Jahres 1999, S. 21-24.

Schmidt-Schütz A, Huss J (1998): Wiederbewaldung von Fichten-Sturmwurfflächen auf vernässten Standorten mit Hilfe von Pionierbaumarten. In: Fischer A (Hrsg.): Die Entwicklung von Wald-Biozönosen nach Sturmwurf. ecomed Verlagsgesellschaft AG & Co. KG, Landsberg, S. 188-211.

Schulze E-D (1989): Air pollution and forest decline in a spruce (*Picea abies*) forest. Science 244(4906): 776-783. 10.1126/science.244.4906.776

Schütz J-P, Götz M, Schmid W, Mandallaz D (2006): Vulnerability of spruce (*Picea abies*) and beech (*Fagus sylvatica*) forest stands to storms and consequences for silviculture. European Journal of Forest Research 125: 291-302.

Seidl R, Schelhaas M-J, Rammer W, Verkerk PJ (2014): Increasing forest disturbances in Europe and their impact on carbon storage. Nature Climate Change 4(9): 806-810.

Skoglund J, Verwijst T (1989): Age structure of woody species populations in relation to seed rain, germination and establishment along the river Dalälven, Sweden. Vegetation 82: 25-34.

Skvortsov AK (1999): Willows of Russia and adjacent countries - Taxonomical and geographical revision, University of Joensuu, Faculty of mathematics and natural sciences report series, Finland, 307 p.

Sperens U (1997): Long-term variation in, and effects of fertiliser addition on, flower, fruit and seed production in the tree *Sorbus aucuparia* (*Rosaceae*). Ecography 20(6): 521-534.

Staaf H, Jonsson M, Olsen L-G (1987): Buried germinative seeds in mature beech forests with different herbaceous vegetation and soil types. Holarctic Ecology 10: 268-277.

Stancioiu PT, O'Hara KL (2006): Regeneration growth in different light environments of mixed species, multiaged, mountainous forests of Romania. European Journal of Forest Research 125: 151-162.

Stiebel H (2003): Frugivorie bei mitteleuropäischen Vögeln. Carl-von-Ossietzki-Universität Oldenburg, Fakultät für Mathematik und Naturwissenschaften, Wilhelmshaven, book, 219 S.

Stoyan D, Wagner S (2001): Estimating the fruit dispersion of anemochorous forest trees. Ecological Modelling 145: 35-47.

Thompson K, Bakker J, Bekker R (1997): The soil seed banks of North West Europe: methodology, density and longevity. Cambridge University Press, Cambridge. 276 p.

ThüringenForst (2013): Wiederbewaldung im Team. URL: http://www.thueringenforst.de/de/forst/aktuelles/wiederbewaldung/. Stand: 08.04.2013, Abruf: 19.12.2013.

ThüringenForst-AöR (2017): 10 Jahre Kyrill - Eine Apokalypse feiert ihr trauriges Jubiläum. Medieninformation 04/2017, 3 S.

TMLFUN (2011): Standortgerechte Baumarten- und Bestandeszieltypenwahl für die Wälder des Freistaates Thüringen auf Grundlage der forstlichen Standortskartierung unter Beachtung des Klimawandels. Erlass vom 19.08.2011. Thüringer Ministerium für Landwirtschaft, Forsten, Umwelt und Naturschutz, 170 S.

TMLNU (2009): Anschreiben: Wiederbewaldung im Team - Folgeinventur im Staatswald. Thüringer Ministerium für Landwirtschaft, Naturschutz und Umwelt, 2 S.

Unseld R, Bauhus J (2012): Energie-Vorwälder - Alternative Bewirtschaftungsformen zur Steigerung der energetisch nutzbaren Biomasse im Wald durch Integration von schnell wachsenden Baumarten. Schriftenreihe Freiburger Forstliche Forschung, Bd. 91, 217 S.

Vilhar U, Roženbergar D, Simončič P, Diaci J (2015): Variation in irradiance, soil features and regeneration patterns in experimental forest canopy gaps. Annals of Forest Science 72: 253-266.

Wagner S, Müller-Using B (1997): Ergebnisse der Buchen-Voranbauversuche im Harz unter besonderer Berücksichtigung der lichtökologischen Verhältnisse. Schriftenreihe, Landesanstalt für Ökologie, Bodenordnung und Forsten/Landesamt für Agrarordnung Nordrhein-Westfalen Nr. 13: 17-30.

Wagner S, Wälder K, Ribbens E, Zeibig A (2004): Directionality in fruit dispersal models for anemochorous forest trees. Ecological Modelling 179: 487-498.

Weber D, Jenssen M (2006): Der Übergang von der Waldnatur zur Forstkultur. In: Fritz P (Hrsg.): Ökologischer Waldumbau in Deutschland. oekom Verlag, München, S. 10-49.

Went M (2011): Die Sukzession auf Sturmwurfflächen des Orkanes „Lothar" in Baden-Württemberg unter besonderer Berücksichtigung der Beräumungsintensität und der Höhenstufe. Technischen Universität Dresden, Fakultät Forst-, Geo- und Hydrowissenschaften, Tharandt, Diplomarbeit, 141 S.

Wohlgemuth T, Kuhn N, Lüscher P, Kull P, Wüthrich H (1995): Vegetation- und Bodendynamik auf rezenten Windwurfflächen in den Schweizer Nordalpen. In:

Schweizerische Forstverein (Hrsg.): Entwicklung von Windwurfflächen in der Schweiz - Erste Forschungsergebnisse, Beobachtungen und Erfahrungen. Schweizerische Zeitschrift für Forstwesen 146: 873-891.

Wohlgemuth T, Kull P, Wüthrich H (2002): Disturbance of microsites and early tree regeneration after windthrow in Swiss mountain forests due to the winter storm Vivian 1990. Forest Snow and Landscape Research 77(1-2): 17-47.

Wohlgemuth T, Nussbaumer A, Burkart A, Moritzi M, Wasem U, Moser B (2016): Muster und treibende Kräfte der Samenproduktion bei Waldbäumen. Schweizerische Zeitschrift für Forstwesen 167(6): 316-324.

Wright SJ, Trakhtenbrot A, Bohrer G, Detto M, Katul GG, Horvitz N, Muller-Landau HC, Jones FA, Nathan R (2008): Understanding strategies for seed dispersal by wind under contrasting atmospheric conditions. PNAS 105(49): 19084-19089.

Zerbe S (2001): On the ecology of *Sorbus aucuparia* (Rosaceae) with special regard to germination, establishment and growth. Polish Botanical Journal 46(2): 229-239.

Zerbe S, Kreyer D (2007): Influence of different forest conversion strategies on ground vegetation and tree regeneration in pine (*Pinus sylvestris* L.) stands: a case study in NE Germany. European Journal of Forest Research 126: 291-301.

Zerbe S (2009): Renaturierung von Waldökosystemen. In: Zerbe S, Wiegleb G (Hrsg.): Renaturierung von Ökosystemen in Mitteleuropa. Spektrum Akademischer Verlag, Heidelberg, S. 153-182.

Żywiec M, Ledwoń M (2008): Spatial and temporal patterns of rowan (*Sorbus aucuparia* L.) regeneration in West Carpathian subalpine spruce forest. Plant Ecology 194: 283-291.

Żywiec M, Holeksa J, Ledwoń M (2012): Population and individual level of masting in a fleshy-fruited tree. Plant Ecology 213: 993-1002.

Żywiec M, Holeksa J, Wesołowska M, Szewczyk J, Zwijacz-Kozica T, Kapusta P, Bruun HH (2013): *Sorbus aucuparia* regeneration in a coarse-grained spruce forest - a landscape scale. Journal of Vegetation Science 24: 735-743.

Chapter 2

Seed dispersal capacity of *Salix caprea* L. assessed by seed trapping and parentage analysis

Katharina Tiebel[1], Ludger Leinemann[2], Bernhard Hosius[2], Robert Schlicht[3], Nico Frischbier[4] & Sven Wagner[1] (2019)

European Journal of Forest Research 138: 495-511. - doi: 10.1007/s10342-019-01186-2

[1]TU Dresden, Institute of Silviculture and Forest Protection, Chair of Silviculture, Pienner Str. 8, 01737 Tharandt, Germany

[2]ISOGEN at the Institute of Genetics and Forest Tree Breeding, Büsgenweg 2, 37077 Göttingen, Germany

[3]TU Dresden, Institute of Forest Growth and Forest Computer Sciences, Chair of Forest Biometrics and Forest Systems Analysis, Pienner Str. 8, 01737 Tharandt, Germany

[4]ThüringenForst, Forestry Research and Competence Center, Jägerstraße 1, 99867 Gotha, Germany

2.2 Introduction

Salix ssp. is a large genus with approximately 330-500 species worldwide (Argus 2006; Dickmann and Kuzovkina 2014), and a high ecological relevance within forest ecosystems (Barnes et al. 1998; Richardson et al. 2014). Willows are known to enhance the soil nutrient status, to decontaminate soils via phytoextraction, to rapidly regenerate and colonize damaged forest areas, to prevent erosion of exposed soil, e.g., on disturbed forest sites, to act as struc-tural elements with a long-term stabilizing effect, and to provide habitats and food for many organisms (see Horvat-Marolt 1974; Kay 1985; Kuzovkina and Quigley 2005; Argus 2006; Baum et al. 2009; Regvar et al. 2010).

Among the most important physiological and autoecological characteristics of willow species are drought tolerance, adaptation to open-site light regimes, and high tolerance to low nutrient availability, which enable willows to colonize and restore disturbed sites. Together with other pioneer tree species, the genus is able to initiate pioneer communities on disturbed sites and to thus re-establish a forest ecosystem (Argus 2006; Kuzovkina and Quigley 2005). At disturbed sites, the surface shading provided by pioneer trees recreates forest climate conditions for subsequent climax tree species which require shelter for successful establishment (Kuzovkina and Quigley 2005; Zerbe 2009). Due to their low site requirements and their ability to withstand extreme climatic conditions, willow species are also important colonists in forest areas with altered disturbance regimes due to climate change.

Willow species may be also cultivated in plantations to produce biomass for bioenergy production (Kuzovkina and Quigley 2005), and some species like *Salix alba* are even used for timber and veneer production (Schirmer 2006). Goat willow (*Salix caprea*) is the most important willow species in European temperate and boreal forests, due to its moderate drought tolerance and its ability to persist to some degree in the forest understory (Skvortsov 1999; Dörken 2011; Perdereau et al. 2014). *S. caprea* is one of the few admixed tree species found within managed spruce forests at high altitudes (Schütt 2006; Perdereau et al. 2014). Early flowering, dioecious *S. caprea* is pollinated by wind and insects, with wind accounting for up to 50 % of pollination (Vroege and Stelleman 1990; Füssel 2007; Dötterl et al. 2014). The small seeds (1-1.5 mm) with cotton hairs (i.e., 'plumed') are wind dispersed (i.e., 'anemochorous') over long distances in May and June (Brouwer and Stählin 1975; Chmelar and Meusel 1986; Skvortsov 1999).

The seed dispersal of anemochorous and zoochorous tree species has been investigated in numerous studies (see McVean 1953, 1956; Matlack 1989; Wagner 1997; Houle 1998; Karlsson 2001; Wagner et al. 2004; Gage and Cooper 2005; Żywiec and Ledwoń 2008; Huth 2009; Schmiedel et al. 2013; Żywiec et al. 2013). However, 'despite their worldwide distribution and ecological importance, very little research has been conducted on the seed and seedbed ecology of the willow species' (Young and Clements 2003). Zasada and Densmore (1979), Niiyama (1990), Leder (1992), Sacchi and Price (1992), van Splunder et al. (1995), Küßner (1997) and Chantal and Granström (2007) studied patterns in established seedlings of willow species. However, such studies of established young trees are limited in terms of the precise information they can provide about seed dispersal distances. This lack of information about seed dispersal distances in regeneration studies can be overcome by using methods like seed trapping and distance to seed source or parentage analyses (Nathan and Muller-Landau 2000). Numerous studies have highlighted the enormous dispersal potential of willow seeds (see Ryvarden 1971; Brouwer and Stählin 1975; Lautenschlager 1994; Barsoum 2002; Gage and

Cooper 2005; Kuzovkina and Quigley 2005; Argus 2006; Seiwa et al. 2008). However, there is only very little specific information on the dispersal distances of willow species. Gage and Cooper (2005) reported that more than 90 % of the seeds of a willow community were deposited within 200 m of the seed trees. But at a distance of 1500 m from the seed trees, the authors still recorded a seed density of 20-30 n m⁻². Ryvarden (1971) assumed shorter dispersal distances for willow, as 93 % of all seeds were caught within 5 m of the seed source.

Generally, dispersal distance is influenced by many factors, such as seed morphology and weight, vertical and horizontal wind direction, wind speed, turbulence, release height and velocity, as well as relief and vegetation cover (Okubo and Levin 1989; Skarpaas et al. 2006; Seiwa et al. 2008). Secondary dispersal by wind and water is likely also important for the transport of willow seeds (Gage and Cooper 2005; Seiwa et al. 2008; Boland 2014). More than 50 % of willow seeds deposited on dry sand were transported a second time by wind (Gage and Cooper 2005). To date, there is insufficient empirical information on seed dispersal distances and the seed tree densities required for the successful regeneration of *S. caprea* on disturbed forest sites.

Genetic parentage analyses provide insights into gene flow via pollen or seeds of a specific parent tree population (Cortés et al. 2014). Perdereau et al. (2014) recorded seed and pollen dispersal over more than 200 km between *S. caprea* populations in Ireland by applying genotyping, which suggested no barriers to gene flow as a consequence of human landscape fragmentation at this scale. A study of chloroplast DNA variation by Palmé et al. (2003) also reported no significant genetic diversity for *S. caprea* populations in Europe because of high willow seed and pollen dispersal distances.

In order to assess the natural regeneration potential of *S. caprea* seedlings on windthrown forest sites, information about the temporal and spatial patterns of seed rain is needed. Up to now, it has been assumed that the abundance and the spatial pattern of parent trees in the area surrounding a particular disturbed site are important determinants upon which forest managers could base silvicultural decisions in relation to the desired natural regeneration concept. To confirm or reject this assumption, we therefore studied seed dispersal of *S. caprea* in 2015 and 2016 using seed traps and performed parentage analyses of established saplings on windthrown sites in Thuringia, Germany. There were four hypotheses guiding the study. (1) Willows do not have a high seed crop every year. We assume that willow seed production has an interannual variability similar to mass-seeding trees. (2) The start of the seed rain is influenced by weather conditions, whereas the length of the seed rain period shows interannual variations. (3) Willow seeds are dispersed by wind over long distances, but the seed density depends on the distance to the seed source, decreasing rapidly with increasing distance. (4) However, the number of deposited willow seeds is also influenced by the number of present seed trees, relief inclination and directionality. We assume that a group of seed trees may produce overlapping seed shadows, resulting in higher seed numbers than in the vicinity of a single seed tree. Downhill seed dispersal by wind will increase dispersal distance, while uphill dispersal will decrease distances. Therefore, a directionality of the seed shadows due to wind should be obvious in the studied species (see Huth 2009; Stoyan and Wagner 2001).

2.3 Materials and methods

2.3.1 Study area

The study area is located at high elevations and along the ridges of the Thuringian Forest, a mountain range in the federal state of Thuringia, Germany (50°40′N and 10°45′E). It is situated between 400 and 982 m above sea level (a.s.l.), with a prevailing south-westerly exposition. The area is characterized by many slopes and an almost total absence of plateaus (Burse et al. 1997; Waesch 2003; Gauer and Aldinger 2005). The mean annual precipitation ranges from 800 mm in the south-west to 1200 mm along the ridges and falls to a level of 700 mm in the north-east (Burse et al. 1997; Gauer and Aldinger 2005; Bushart and Suck 2008). The annual average temperature in the region varies between 4 and 6 °C (Burse et al. 1997; Bushart and Suck 2008). The area is influenced by an Atlantic, moderately cool and moist central mountain climate (Burse et al. 1997; Gauer and Aldinger 2005). The prevailing winds are from the southwest, with a secondary wind maximum originating from the north-east. The average annual wind speed in the study area is 3.5-4.5 m s^{-1} (Bürger 2003). The dominant soil types of the forest sites are low-base cambisols with low to medium nutrient contents (Gauer and Aldinger 2005). The landscape features a largely contiguous forest system, with ~ 90 % forest cover, some small upland meadows in stream valleys and occasional small raised bogs. The study area is dominated by single-layered, even-aged Norway spruce forests (*Picea abies* (L.) Karst.), although *Luzulo-Fagetum* and *Asperulo-Fagetum* beech forests are the predominant potential natural vegetation types (Frischbier et al. 2014).

We selected five study sites (A-E) located on slopes and mountain tops (plateaus) at higher elevations and near the ridges of the Thuringian Forest (715-900 m a.s.l.). The choice of sites made it possible to investigate uphill, downhill and level seed dispersal. Each site consisted of an open area surrounded by a forested search zone of 500 m. The open areas were windthrown by the cyclone 'Kyrill' in January 2007 (Fink et al. 2009). Before the storm, the sites were dominated by 68-100-year-old spruce stands. After the storm, the damaged areas were completely cleared, and no willow seed trees were present in the open areas. The size of the open areas ranged from 4-12.7 ha (Table 2.1), and no closed regeneration layer had yet established at any of the study sites. The open areas were surrounded by 59-122-year-old spruce forests admixed with a small number of isolated *Betula pendula* Roth, *Salix caprea* L. and *Sorbus aucuparia* L. The meteorological data for the seed trapping periods are listed in Table 2.2.

Table 2.1 Study site data (*DBH* diameter at breast high, *parent trees* female and male willow trees, *seed trees* only female willow trees, *SD* standard deviation, *relief-induced dispersal* classifications of further investigated study sites by geostatistical models).

Open area at study site	A	B	C	D	E
Relief-induced dispersal	-	'Uphill'	-	'Level'	'Downhill'
Elevation above sea level (m)	845-900	735-765	840-880	865-895	715-775
Topography	Mountain peak with one slope	Mountain peak with slopes	Flat area	Mountain peak with slopes	Slopes
Size of open area (ha)	5.98	4.03	7.46	5.59	12.70
Seed traps (n)	27	26	38	34	41
Female/male trees within 500 m forested search zone around the open area (n)	2/4	15/7	0	11/9	126/130
Minimum distance between seed tree and seed trap (m)	385	504	-	289	76
Dbh ± SD of parent trees (cm)	16.2 ± 4.6	19.8 ± 9.8	-	16.5 ± 6.4	16.5 ± 4.3
Dbh ± SD of seed trees (cm)	13.5 ± 3.5	16.1 ± 6.3	-	17.5 ± 7.5	17.4 ± 4.6

Table 2.2 Meteorological data (half-hourly values; climate station 'Grosser Eisenberg' - 50° 37′ 24″ N and 10° 46′ 59″ O) of the seed trapping periods in 2015 (3 months) and 2016 (1.5 months) in the study area (*SD* standard deviation).

	2015				2016	
	From mid-Apr	May	Jun	Until mid-Jul	From mid-May	Jun
Wind spead (m/s)						
Minimum	0.9	0	0	0	0	0.5
Maximum	7.2	7.6	9.2	7.2	5.0	5.3
Mean	3.3	3.2	3.3	3.1	2.2	2.2
SD	1.2	1.5	1.4	1.2	0.7	0.9
Wind direction (°)						
Mean	161	208	185	234	189	190
Median	217	236	234	252	223	237
SD	109.6	95.6	102.6	71.2	95.2	92.6
Mean temperature (°C)	7.5	10,1	13.4	18.5	11.6	14.6
Precipitation (mm/month)	19.9	35.1	96.0	68.9	58.5	129.0

2.3.2 Experimental design

Within the 500-m forested search zone at each study site, we mapped all *S. caprea* trees that were expected to potentially produce seeds (≥ 5 cm diameter at breast high [dbh]), using a blumax Bluetooth GPS-4013 Receiver. For each *S. caprea* tree, we recorded the dbh and determined its gender during flowering season. The search zone distance of 500 m was chosen as a compromise between feasibility and prior knowledge of goat willow seed dispersal (Gage and Cooper 2005). Except for site C, 2-126 female (i.e., seed) trees were identified at each study site, mostly located along forest edges or roadsides. At site C, no seed trees were found within the 500 m search zone. Although the search distance was expanded to 900 m at this site, no seed trees were identified.

At all study sites, 26-41 seed traps were placed within the open areas. Due to their vast areal extent, seed traps were placed along two crossing line transects with intervals of 20 m between the traps, rather than along a regular grid (Fig. 2.1a - see also Bjorkbom 1971; Greene and Johnson 1996). The orientation and length of the line transects were not uniform, due to the differences in the size and shape of the open areas. The line transects extended over the entire open area of each study site and into the surrounding spruce forests. The minimum distances between the seed trees and the nearest seed trap ranged between 76 and 504 m (Table 2.1).

In order to gain knowledge about seed dispersal close to seed trees in the forest, a total of 63 additional seed traps were placed around three selected individual goat willow seed trees (b, d, and e) located along forest roads in or near the search zone at study sites B, D and E, respectively (504, 554 and 276 m). The selected trees b, d, and e had a height of 13.5, 8.7 and 13.7 m, and a dbh of 22.1, 27.9 and 25.9 cm, respectively. Seed traps were place along two or three line transects extending along forest roads for up to 50 m away from these trees (Fig. 2.1b).

The deployed seed traps featured a sticky, non-drying substance (glue product: 'Raupenleim', © 2017 F. Schacht GmbH & Co. KG) covering a surface area of 0.043 m² (see also Gage and Cooper 2005; Kollmann and Goetze 1998; Werner 1975). The glue was painted onto standardized transparent foils, which were wrapped around vertical cylinders to allow for analysis of both seed shadow and dispersal direction (Fig. 2.1c). The cylinders were fixed onto a bar 1.5 m above the ground. The traps were checked every 3 or 4 weeks, when the foils were replaced. The seeds on the foils were then counted in the laboratory. Seed dispersal in the open areas was studied in 2015 and 2016 for 3 months (mid-April to mid-July) and 1.5 months (mid-May to end of June), respectively. Seed dispersal around the single trees in the forested zone was only studied in 2016 (mid-May to end of June). In 2015, the seed rain produced by goat willow seed trees located in the lowlands (at approximately 450 m a.s.l.), well outside the study area, started a month earlier than at the high-altitude study sites. In 2016, the seed rain took place simultaneously at all altitudes.

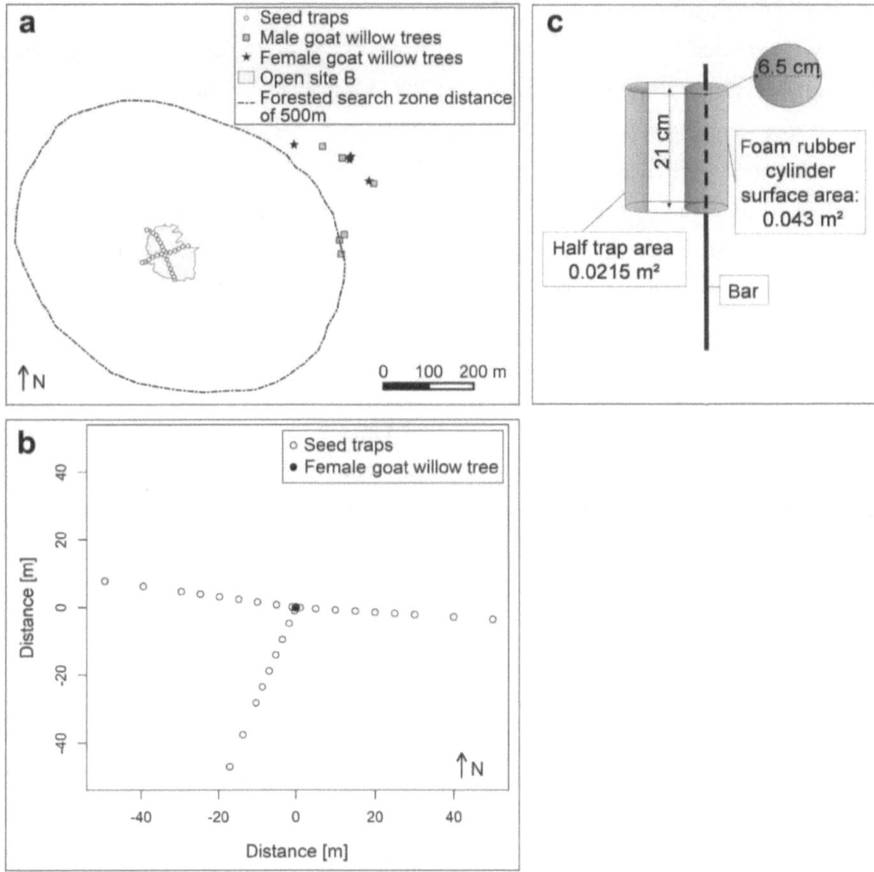

Fig. 2.1 a Example of the experimental study design in the open areas, e.g., study site B. **b** Example of the experimental study design around individual *Salix caprea* seed trees located in the forested search zone, e.g., seed tree b. **c** Technical specifications of the sticky seed traps.

2.3.3 Genetic analysis

At the study site D, the gene flow was estimated using genetic markers. Genetic analyses of established saplings provided supporting information relating to seed dispersal. All potential parent trees within the 500 m forested search zone around the open area at study site D were identified and genotyped. Twigs were sampled from 100 *S. caprea* saplings chosen randomly within the open area and from all 20 parent trees (9 males and 11 females) during the spring of 2016. Sapling positions were recorded using GPS. Their age was determined by counting the growth rings at the stem base. Sapling ages varied between 2 and 9 years, with a mean age of 5 years. All analysed willow saplings had established after the cyclone Kyrill (2007). The age of the parent trees was unknown.

DNA was extracted from frozen bud tissue using DNeasy96-Kit (Qiagen, Hilden). Different nuclear-DNAprimers were tested (Barker et al. 2003; Hanley et al. 2006), and finally, seven nuclear loci were genotyped: SB880, SB24, SB38, SB49, SB80, Sa54B and Cha475. The fragment lengths were analysed with an ABI-3110-capillary sequencer. The number of different alleles (Na) and the genetic diversity (Ne, Eq. 2.1) were calculated per locus and sample size (n) of the parent tree and offspring population using GenAlEx 6.5 (Peakall and Smouse 2012). The formula used to determine the genetic diversity was:

$$N_e = \frac{1}{\sum_{i=1}^{k} p_i^2}$$
(2.1)

with p_i being the frequency of the $i = 1,\dots\ k$ alleles at a locus in a population (Nei 1978). In the sapling population, Na and Ne of one individual could not be determined in almost all loci (see Table 2.5). The genetic distance (D, Eq. 2.2) of the accumulated allele differences per locus between the two populations was calculated as follows:

$$D = -log_e I$$
(2.2)

with I being the normalized gene identity between the populations (Nei 1972, 1978). Parentage was assessed by simple exclusion based on multilocus genotypes. Exclusion was calculated using the program 'Genfluss' (Leinemann not publ.). This technique uses differences between potential parents and offspring to reject parentage for a specific sapling. Trees lacking a given allele can readily be excluded as potential parents. Whenever a sapling was identified as offspring of a particular female goat willow tree, the distance between the parent tree and the sapling was measured and recorded as the respective seed dispersal distance.

2.3.4 Seed trap data analysis

As seed count data were non-normally distributed but approximately negative-binomially distributed, differences in seed numbers among the different study sites and between the two study years were analysed using the Kruskal–Wallis *H*-test. Where significant differences were ascertained ($p < 0.05$), the Mann–Whitney U-test was applied with a Bonferroni correction as an adjustment method to obtain additional information about the groups of differences (Zar 2010). Significant differences were accepted at $p < 0.05$. Statistical analyses were conducted using the statistics software R (version 3.3.2). Quantum GIS (QGIS 2.4.0 Chugiak) was used to create maps of all study sites based on original forest maps and aerial orthophotographs. These maps were used to outline the boundaries of the open areas at each study site, to determine the surrounding 500 m forested search zone, and to mark the positions of parent trees, sampled willow saplings, and seed traps.

2.3.5 Geostatistical models

Model-based geostatistics were applied to test for the effects of distance to goat willow seed trees on trapped seed numbers. Inverse models are well-developed tools for the analysis of horizontally oriented seed traps (Clark et al. 1999; Ribbens et al. 1994; Stoyan and Wagner 2001). However, inverse models cannot be applied to data obtained from seed traps with a vertically oriented trap surface. Thus, we chose a flexible geostatistical approach which allows handling spatially auto-correlated data. Regrettably, geostatistical models do not provide the seed dispersal kernel of a single seed tree, because these models can only consider the nearest seed tree as influential to data from individual seed traps. Like also reported by Gage and Cooper (2005), Leder (1992) and Ryvarden (1971), our model does not explicitly account for overlapping seed shadows of multiple seed trees in an area.

Within R (R Core Team 2014), we used the 'geostatsp' package (version 1.7.4; Brown 2015) to fit non-Gaussian models using the INLA procedure (version 1.7.4). INLA performs a wide range of Bayesian statistical analyses by applying sophisticated approximations for handling the numerical difficulties that commonly arise in this context (Rue et al. 2009). Details of the statistical models are given in Table 2.3.

Starting from suitable prior distributions, we obtained approximate posterior distributions and credible intervals (Robert 2007, Sect. 5.5). In analogy to confidence intervals, a parameter is considered significant if the 0.95-credible interval with endpoints defined by the 0.025th and 0.975th quantiles of the posterior distribution does not include 0. However, the interpretation differs from confidence intervals: it does not refer to the hypothetical distribution assuming the parameter is exactly 0, which typically has zero probability under the prior; rather, the deviations manifesting themselves in the data are significant enough to state that there is at least a 97.5 % chance of drawing the correct conclusion about whether the parameter is greater or less than 0.

Table 2.3 Specification of the Bayesian geostatistical model: assumptions regarding prior distributions and the distribution of the data.

	Parameter/data	Distribution
Hyperparameters	Range ϕ	Gamma with 0.025th and 0.975th quantiles at 20 and 100
	Precision $1/\sigma^2$	Gamma such that σ has a distribution with 0.025th and 0.975th quantiles at 0.1 and 3.0
	Size (shape) r	'PC prior' such that the square root of 2 times the KL divergence of a mean 1 gamma distribution with shape r from one with shape approaching ∞, with suitable asymptotic scaling, has an exponential distribution with rate parameter 7 (Simpson et al. 2017)
Fixed effects	Intercept a	Normal with mean 0 and variance approaching ∞
	Slope b	Normal with mean 0 and variance 1000
Random effects	Spatial field $U(s)$ (where s are the points in the plane)	Joint normal with mean zero and covariances $\sigma^2 \frac{\sqrt{8}d}{\phi} K_1(\frac{\sqrt{8}d}{\phi})$ for points with distance d, where K_1 is a modified Bessel function (Matérn correlation structure)
Observed numbers of seeds	n_i at point s_i for traps $i=1, \ldots, m$	Independent negative binomial with mean $e^{a+b\ \mathrm{distminlg}_i+U(s_i)}$ and size parameter r; equivalently, Poisson with mean $e^{a+b\ \mathrm{distminlg}_i+U(s_i)+V(s_i)}$, where $V(s_i)$ are additional independent noise terms such that $e^{V(s_i)}$ has a mean 1 gamma distribution with shape r (NB2 regression model)

A general question in the Bayesian context is the robustness of the results against possible misspecifications of the priors (Berger 1985, Sect. 4.7), particularly if there is little advance knowledge. In our case, the distribution of the range parameter follows typical distances expected from experience in the field, while the priors for the fixed effects are rather flat and are not expected to dominate the results. The size parameter, which controls overdispersion, has a distribution chosen in order to favour values representing a small overdispersion, and in any case can be seen as part of the spatial random field. The remaining parameter is the standard deviation σ (or the precision $1/\sigma 2$), which appears to be the most critical one. We therefore repeated our computations with a broad range of other plausible priors for σ. While the general patterns remain qualitatively similar, the choice is clearly visible in the corresponding posterior distributions. Sufficient caution is therefore needed when interpreting the statistical results, and, in particular, one should avoid attaching too much meaning to precise numerical values.

The estimation of the range and standard deviation of the random part of the model is also performed during the fitting procedure. Here, the range ϕ gives an exact sense to the conventional concept of the area of influence of a sample (Chilès and Delfiner 1999). It is derived from a variogram and quantifies the distance between two points on a plane beyond which the correlation of values measured at these points, e.g., the number of seeds in a trap, becomes small. The standard deviation σ measures the strength of the spatial random effects.

To create geostatistical models of seed rain, the data sets obtained in 2016 from the open area at study sites B, D and E were combined with the 2016 data sets of the respective individual willow trees b, d and e in the forested zone as site-specific data sets B (= B+b), D (= D+d) and E (= E+e). The data sets were split into the categories 'uphill' (B), 'level' (D) and 'down-hill' (E) on the basis of the relief-related seed dispersal.

2.4 Results

The numbers of trapped seeds at open areas A-E were significantly higher in 2016 than in 2015 (Mann–Whitney U-test: $p = 0.003$). Average seed numbers at the five open areas ranged between 0.6 and 1.8 n per trap in 2015 and 1.1-2.1 n per trap in 2016 (Fig. 2.2).

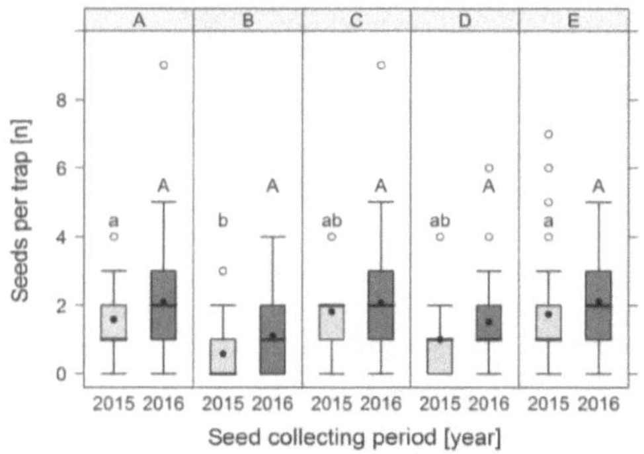

Fig. 2.2 *Salix caprea* seed numbers [n per trap] collected in 2015 and 2016 in the open areas. Lower and upper case letters indicate significant differences among study sites (A-E) in 2015 and 2016, respectively (KruskalWallis *H*-test: $p < 0.05$). White circles show outliers and black circles inside the boxes are mean values.

In both years, the highest mean seed numbers occurred at open area E, next to a stand consisting of *S. caprea*, *Sorbus aucuparia* and *Betula pendula*, as well as at open area C, where no seed trees were found within the extended 900 m forested search zone. The lowest seed rain over the 2-year study period was observed for the open area at study site B, located more than 504 m from the nearest seed source (Table 2.1). The comparison of all open areas showed no significant differences of deposited seed numbers in 2016 (Kruskal-Wallis *H*-test: $p > 0.05$). In 2015, open area B differed significantly from open area A and E (pairwise Mann–Whitney U-test: $p < 0.05$).

2.4.1 Temporal patterns of seed dispersal

In 2015, seeds were trapped for a period of three months, from mid-April to mid-July. In the first month, only goat willow seeds produced by seed trees in lowland areas outside the study area were caught. On-site observation revealed that catkins at the high-altitude sites and on the ridges were still closed at this time. One month later, the goat willow seed rain had also started at the higher altitudes, including the seed trees at the study sites. With the exception of site E, the highest percentage of seeds were deposited in the seed traps (65-97 %) during the first month of the 2015 collection period (early = mid-April to mid-May). The seed rain continued from mid-May until the end of the collection period in mid-July, but the relative proportion of deposited seeds decreased to 0-35 % (Fig. 2.3).

In 2016, the seed rain took place simultaneously at all altitudes over a period of 6 weeks from mid-May to the end of June. The vast majority of all seeds (92-100 %) was trapped in the first part of this period (early = mid-May to early-June). At open areas A and E, there was no measurable seed rain in the remainder of the collection period (late = early-June to the end of June).

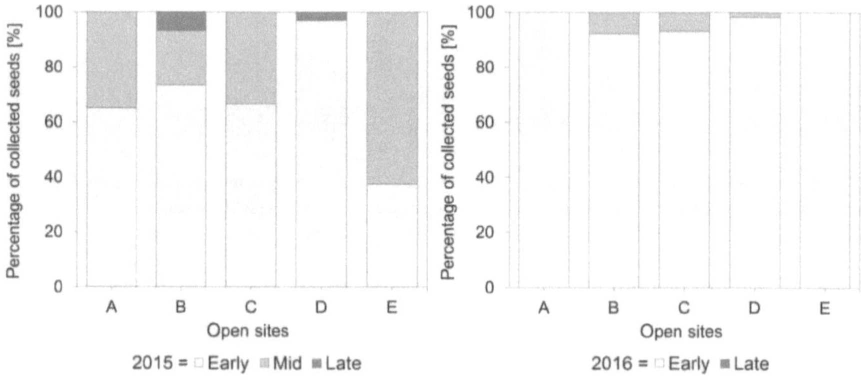

Fig. 2.3 Percentage of *Salix caprea* seeds collected in the open areas at study sites A-E in 2015 (left) and 2016 (right) during each 4-weekly collection period (2015 = early: mid-April to mid-May, mid: mid-May to mid-June, late: mid-June to mid-July; 2016 = early: mid-May to early-June, late: early-June to end of June).

2.4.2 Dispersal distance and spatial patterns of seed dispersal

In this section, trapped seed numbers are presented by distance to the nearest seed tree, and overlapping seed shadows may occur in all traps. For the individual willow seed trees located in the forested search zone, maximum seed numbers of 23 (tree e), 106 (tree b) and 156 n per trap (tree d) were recorded in 2016 close to the stem base and underneath the tree crowns

(Fig. 2.4b-d). No visibly differing trends were observed for transects oriented in different directions. Seed rain distribution and trapped seed numbers differed among individual willow trees, with seed numbers generally decreasing rapidly with increasing distance from the respective seed source. On average, 2.3 seeds per trap were caught at 40-50 m distance from the seed source in the forested zone. At the open area of site B, which featured the largest distances between seed source and traps, 0-4 seeds per trap (average of 1.1 seeds per trap) were recorded 700-870 m from the seed sources. The seed rain around the individual goat willow trees in the forest and in the open areas resembles a graph of a negative exponential function with a steep slope (Fig. 2.4a).

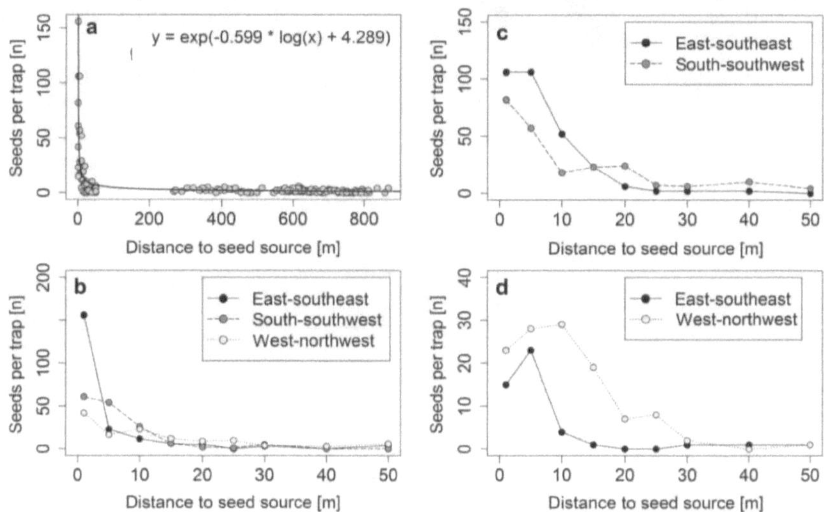

Fig. 2.4 a Combined 2016 dataset of trapped *Salix caprea* seed numbers [n] around individual seed trees b, d and e in the forested search zone and in the open areas at study sites B, D and E plotted against distance to seed source [m]. **b-d** *S. caprea* seed numbers per trap [n] around the individual seed trees b, d and e in the forested search zone in 2016 depending on seed dispersal direction and plotted against distance to seed source [m]. Note the different *y*-axis scales for all plots and the *x*-axis in Fig. 2.4a.

In 2016, *S. caprea* seeds were dispersed at the study sites in a pattern shown in Fig. 2.5a. Farther than 350 m from the respective seed source, no differences in deposited seed numbers with respect to the number of seed sources (isolated seed trees at study sites B and D vs. stand of seed trees near study site E), relief inclination (study site B-'uphill', study site D-'level' and study site E-'downhill') or directionality (east, south and west) were observed (Fig. 2.4 and 2.5). The associated maps of the seed shadows estimated using geostatistical models are shown in Fig. 2.5b. The geostatistical model predictions for the three data sets were significant for the expected values of logarithmic distance to the nearest seed trap and trapped willow seed numbers (see Table 2.4). Models revealed only slightly differing spatial patterns for estimated uphill (site B), level (site D) and downhill (site E) dispersal (Fig. 2.5). The goodness of fit of the models can be assessed by the correlation of the measured data and the graph of the negative exponential function which was used for geostatistical modelling in Fig. 2.4a. While Fig. 2.5b shows modelled maximal goat willow seed dispersal distances of 500-800 m, the measured data sets featured no dispersal limit up to 870 m from the seed source (Fig. 2.5a).

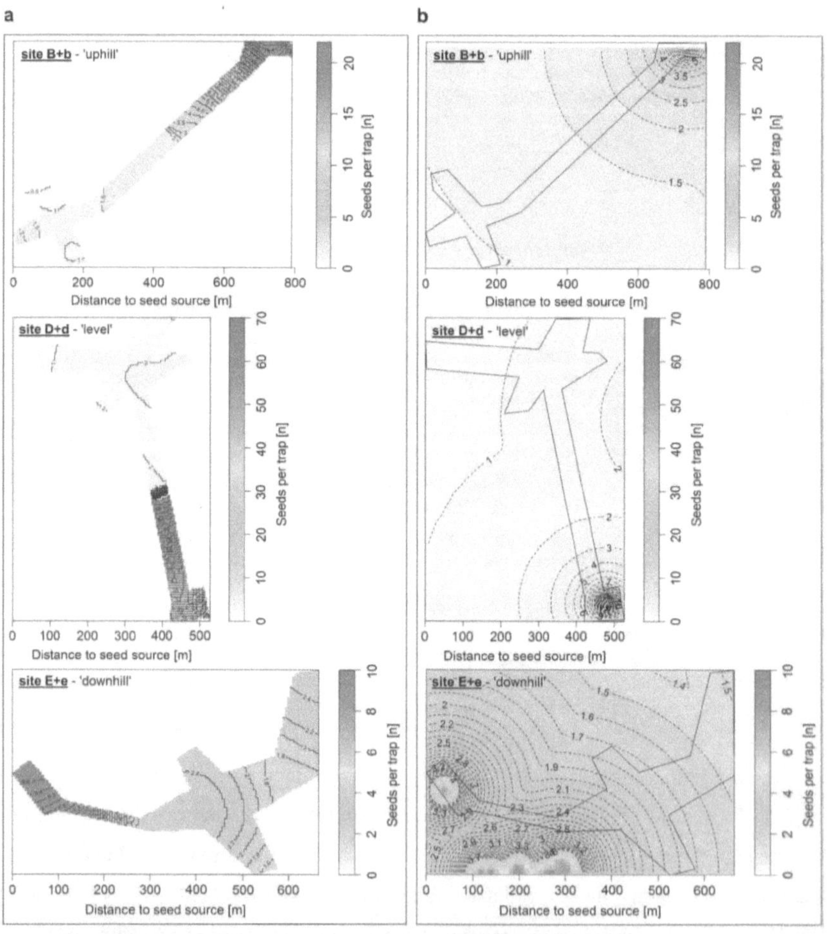

Fig. 2.5 a Interpolated distribution of *Salix caprea* seeds [n] trapped in 2016 for the site-specific data sets B + b, D + d and E + e. **b** *Salix caprea* seeds per trap [n] predicted by geostatistical models for the site-specific data sets B + b, D + d and E + e. The isolines in **a** and **b** represent smoothed values from 53 (B + b), 52 (D + d) and 59 (E + e) trap positions. Note the different site-specific scales of distance (x-axis) and seed numbers (y-axis).

Table 2.4 Results of geostatistical negative exponential dispersal model of *Salix caprea* seeds for 2016: characteristics of the posterior distributions (*SD* standard deviation).

Data set of study site		Mean	SD	0.025th quantile	0.975th quantile
B	Intercept a	3.67	0.96	1.75	5.59
	Slope b	-0.58	0.17	-0.92	-0.24
	Range ϕ	44.02	15.59	20.82	81.40
	Standard deviation σ	0.92	-	0.55	1.52
	Size r	3.57	1.59	1.43	7.54
D	Intercept a	5.57	0.93	3.76	7.48
	Slope b	-0.91	0.17	-1.27	-0.58
	Range ϕ	41.86	15.46	18.52	78.41
	Standard deviation σ	0.63	-	0.35	1.06
	Size r	5.63	2.56	2.16	12.03
E	Intercept a	2.66	0.74	1.16	4.10
	Slope b	-0.39	0.14	-0.67	-0.11
	Range ϕ	49.39	15.89	25.14	87.02
	Standard deviation σ	0.75	-	0.47	1.20
	Size r	35.48	62.82	3.85	174.73

2.4.3 Genetic parentage analysis

For 29 of the 100 goat willow saplings analysed at study site D, a specific parent tree was successfully assigned from the group of potential parent trees. It was possible to identify 3 of the 11 female and 8 of the 9 male goat willow trees as parents. The assigned offspring samples were evenly distributed, without detectable spatial-genetic variations (Fig. 2.6). A minimum of 71 % of the sapling population originated from parent trees located outside of the 500 m search zone. Only 4 of the aforementioned 29 saplings originated from a pairing of a seed and pollen parent located within the study site and the corresponding search zone. An additional 10 saplings were assigned to a seed parent and 15 saplings to a pollen parent, suggesting an external gene flow via seed for 86 % of the sampled saplings and via pollen for 81 %. The closest female and male parents were located 240 m and 280 m from the nearest edge of site D, respectively. The seed dispersal distance of the most successful seed parent (with 8 offspring identified) was between 550 and 800 m. The age of the offspring ranged between 2 and 9 years, with a large number of 4-6-year-old saplings.

All 100 + 20 sampled individuals had a unique multilocus genotype. The estimated genetic variation of the parent tree and sapling populations revealed a higher allelic variation in the regenerated population (average 16 alleles per locus). The markers exhibited a range of 9 alleles per locus of parent population. The variation was particularly high in the locus SB349, where 11 alleles were observed for parent trees and 25 alleles for saplings. Altogether 63 % of the sampled saplings had so-called private-alleles, which occurred only in the offspring population. Upon comparison of the genetic diversity values of the populations, the locus SB349 revealed quite similar values for both parents and offspring. The genetic distance was 3.8 %. Overall, genetic diversity within the sapling population was higher than in the parent population (Table 2.5).

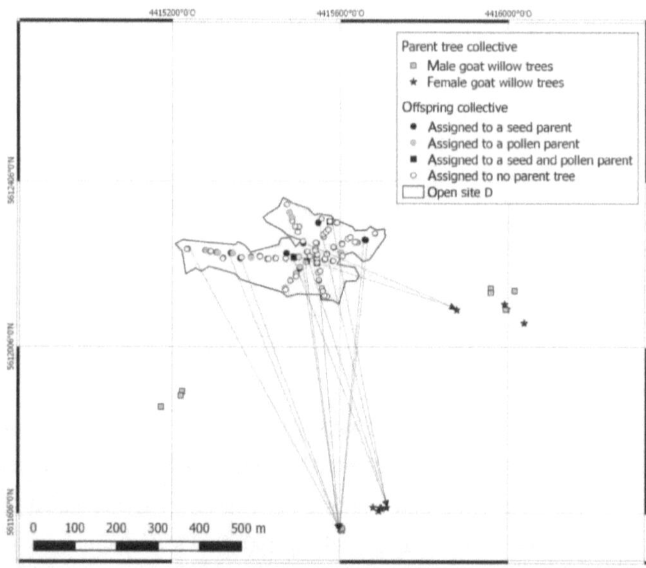

Fig. 2.6 Offspring of the localized *Salix caprea* parent tree population within the 500 m forested search zone at the study site D. Only saplings that were assigned to a specific seed parent are connected to the corresponding parent by an arrow.

Table 2.5 Number of different alleles (*N*a), genetic diversity (*N*e = $1/\Sigma p_i^2$) per locus, and sample size (*n*) of the parent tree and offspring population at study site D.

Loci	Parent trees			Offspring		
	n	**N_a**	**N_e**	**n**	**N_a**	**N_e**
SB880	20	3	1.054	99	4	1.489
SB24	20	10	3.980	100	17	4.373
SB38	20	13	8.889	99	19	9.479
SB349	20	11	5.128	99	25	5.064
SB80	20	12	8.511	99	17	8.751
Sa458	20	13	8.247	99	21	7.675
Cha475	20	4	2.524	99	7	2.625

2.5 Discussion

2.5.1 Seed production and temporal patterns of seed dispersal

The results revealed a high variability in seed production between the two studied years, as assumed in the first hypothesis. Herrera et al. (1998) described that intermittent large and small seed crops are highly frequent in woody species. Pioneer tree species should therefore also be considered mast species like classically known mast-fruiting trees such as beech or oak. The phenomenon of mast-fruiting in tree species is predominantly observed in wind-pollinated species, in environments with changing resource availability, variable weather conditions, and a high density of seed predators (Kelly 1994; Herrera et al. 1998; Kelly and Sork 2002). In many cases, it is difficult to determine a single factor responsible for triggering a mast year, but Kelly et al. (2002) mentioned weather as the most frequent and important factor. Bastide and van Vredenburch (1970) and Gage and Cooper (2005) also recorded differences in trapped willow seed numbers of up to 50 % between seed years due to weather conditions. Frost events in spring may lead to reduced seed production and germination capacity of willow seeds (Young and Clements 2003; Gage and Cooper 2005).

At the level of the individual tree, seed production is also generally influenced by the vitality and dimension of the seed tree; for example, by its crown radius (Fischer et al. 2016). In a variety of willow species (*S. alba*, *S. daphnoides*, *S. elaeagnos*, *S. triandra* and *S. purpurea*), even small individuals of only 2-3 m height may produce 22,000-740,000 seeds in a particular year (Karrenberg et al. 2002; Karrenberg and Suter 2003). The measured data and the geostatistical models in this study also showed a high variability of the number of seeds deposited close to the individual willow seed trees, likely owing to tree-related variation in seed production.

The differences observed in 2015 between goat willows located in lowland areas (450 m a.s.l.) and at high altitudes (> 715 m a.s.l.) with respect to the start and duration of the seed rain (see hypothesis 2) may result from altitudeinduced climatic variability (Kolodziej and Frühauf 2008; Scheffler and Frühauf 2011). Densmore and Zasada (1983) found the phenomenon of altitude-induced variation of the period of seed maturation to be common in Alaskan willow species. In the case of *S. caprea*, the onset of the seed rain is delayed by 2.51 ± 0.16 days per 100 m altitude (Ziello et al. 2009). The seed trees at the study sites would hence have been expected to fructify already 10-13 days after the goat willows located in lowland areas. However, temperature also exerts a strong influence on flowering time in willow species, with cold temperatures delaying flowering. An early warm period in spring (as was the case at the Thuringian sites in 2015) may thus lead to an earlier onset of flowering in willow species compared to years with a long winter and a cold spring (as recorded for the study sites in 2016) (Mosseler and Papadopol 1989). Willow populations being late to release seeds in a particular year do not necessarily also release seeds late in succeeding years (Densmore and Zasada 1983). Mosseler and Papadopol (1989) also observed an influence of spring temperature on the length of the flowering period in Canadian willows. Warmer temperatures resulted in extended flowering duration, as we also observed in our study.

Air humidity and the extent of wind turbulence are other important factors in the seed release process. In willows, mature capsules will only open during periods of low air humidity (Kohlermann 1950). Strong wind turbulence will increase the amount of seed released from capsules compared to high wind velocities without turbulence (Skarpaas et al. 2006). The environmental conditions air humidity, wind turbulence, temperature, wind speed and wind direction, which influence seed release (Kohlermann 1950; Sarvas 1952; Skarpaas et al. 2006; Huth 2009), are subject to an extreme temporal variability. Therefore, seeds which are released over longer periods may be subjected to a greater variation in environmental conditions than seeds released over shorter periods. Huth (2009), for example, found significant month-specific differences in wind direction and speed, which were reflected in the seed shadow of *B. pendula*. Thus, the interannual variability of the start and duration of willow seed rain may lead to varying spatial patterns of the seed shadow (Houle 1998; Nathan and Muller-Landau 2000). Local spatial patterns of seed rain can also differ between mast and non-mast years (Houle 1998).

2.5.2 Dispersal distance and spatial patterns of seed dispersal

Dispersal distances appear to be greater for wind-dispersed willow seeds featuring a pappus than for species with winged seeds or seeds dispersed by birds (see McVean, 1953, 1956; Perala and Alm, 1990; Karlsson, 2001; Huth 2009). The spatial pattern of the goat willow seed rain in this study exhibited a negative exponential distribution in relation to dispersal distance, as previously described by Greene and Johnson (1996) and Hughes and Fahey (1988) for tree seeds with anemochorous dispersal. Small peaks of more than 25 seeds per trap were observed within 10 m of individual seed trees (see Fig. 2.4b-d), and the highest seed numbers of up to 156 n per trap were observed underneath the tree crowns. A distance frequency distribution resembling a leptokurtic pattern was also observed by Gage and Cooper (2005), who reported seed numbers of 200-10,000 n m^{-2} close to American willow species. The same pattern was found by Ryvarden (1971), who trapped 1,600 seeds m^{-2} around alpine willow species.

Like in all aforementioned studies, overlapping seed shadows were also a problem in our study.

Seed bulks and ripe catkins are likely the reasons for the significant fit of the negative exponential function of our geostatistical model and for the recorded seed number peaks close to seed trees. Our observations revealed large numbers of fallen catkins under tree crowns, with capsules either open or still closed. Ryvarden (1971) also observed large numbers of catkins directly beneath trees, with only 7 % of all trapped seeds dispersed more than 5 m from the seed source due to their occurrence as seed bulks. The fine hairs of willow seeds cause seeds to intertwine in the capsules, which are then mostly released as 'seed bulks'. Once in the air, the seed bulks may disintegrate into smaller units or single seeds, unless the seeds are previously deposited on the ground or caught up in vegetation (Kohlermann 1950; Ryvarden 1971; Karrenberg and Suter 2003). The sinking rate of *S. viminalis* seed bulks is 3.8 s m^{-1}, which is significantly faster than the 9.3 s m^{-1} of single seeds (Kohlermann 1950). This difference in sinking rates explains why many of the willow seed bulks were deposited near seed trees.

Nevertheless, the geostatistical models in the presented study confirmed long dispersal distances of 500-800 m for individual goat willows, which confirms hypothesis 3. In our study, 0-6 seeds per trap were recorded at 50 m, and 1-4 seeds per trap at 250 m. Gage and Cooper (2005) measured seed densities of 0-200 n m^{-2} and 0-100 n m^{-2} within 50 m and 200 m of a seed source, respectively. Therefore, the dispersal distances of *S. caprea* do not differ from other willow species. The large seed numbers of 0-9 n per trap measured at open area C, despite an absence of seed trees within the 900 m search zone, give testimony to the huge seed dispersal capacity of goat willow. The assumption of a very large seed dispersal capacity is supported by the fact that lowland goat willow seeds were trapped at the high-altitude study sites in spring 2015, while the seed rain of local seed trees had not yet started. Kohlermann (1950) reported a sinking rate of 7.2 s m^{-1} for *S. caprea*; based on which Schirmer (2006) calculated seed dispersal distances of 2-3 km, even at low wind speeds. Imbert and Lefèvre (2003) reported maximum seed dispersal distances of 1-3 km for a black poplar population whose seeds are morphologically similar to willow.

Secondary drift likely also contributes to such long dispersal distances (Matlack 1989; Gage and Cooper 2005). Depending on soil texture und moisture, up to 50 % of the willow seeds initially deposited on the ground may drift away afterwards due to wind (Gage and Cooper 2005; Seiwa et al. 2008), thus leading to longer dispersal distances. The enormous distances that individual tiny seeds may spread can also be explained by the drift to higher altitudes brought about by convective air currents, as observed, e.g., at the study site B (Lautenschlager

1994; Karrenberg et al. 2002; Karrenberg and Suter 2003). Gage and Cooper (2005) reported a small but constant and, therefore, distance- and direction-independent seed rain, i.e., 'noise', of approximately 10-30 seeds m^{-2}, which reached all of their study sites. Based on our results, we assume a seed source-independent background level of 1.2 goat willow seeds per trap, even if there is only a very small 'background presence' of the species in an area.

The absence of any clear directionality in the seed rain patterns observed around individual seed trees but also in the open areas in relation to seed tree position is unexpected in an anemochorous species (see hypothesis 4). This absence of directionality may be caused by turbulence, secondary drift or seed flow from seed trees located outside the study sites. Kohlermann (1950) referred to lower level, lateral winds of a main wind direction, which can influence dispersal distance and direction regardless of seed mass and sinking rate, even close to seed trees. In open areas, wind profiles have a logarithmic shape with an increase in wind speed with height above ground. Even solitary trees in open areas, and their vertical and horizontal arrangement within the site, may interrupt this logarithmic profile. In addition to temporal effects on wind speed and direction within the observation period, wind profiles thus become more complex because turbulence and variable wind speeds at all heights above ground disturb simple patterns (see Moon et al. 2013).

2.5.3 Genetic parentage analysis

Seed dispersal affects the gene flow, gene structure and diversity of populations and collectives (Barnes et al. 1998; Nathan and Muller-Landau 2000). The low allele coincidences between the parent and the regeneration population as well as the higher number of additional alleles of saplings suggest a significantly larger *S. caprea* parent population than the mapped trees. Goat willow seed trees within and beyond the study site may have contributed to the regeneration. The natural regeneration at study site D took place in the years 2011-2013; the cyclone Kyrill cleared the site in 2007. It is possible that certain parent goat willow trees contributed to the natural regeneration during this time period and then died before our study started in 2015. However, the removal of willow seed trees was prohibited by the public forest owner after the storm event in 2007, and multiple deaths of goat willows due to natural disturbance after the storm seem unlikely. Saplings cannot have originated from the seed bank, as willow seeds remain viable on and in the soil only for short time periods (Junttila 1976; Niiyama 1990; Worrell 1995; Karrenberg and Suter 2003). Taking into account the large dispersal distances of the goat willow pollen and seeds, it thus seems likely that the high number of alleles in the sapling population is due to external gene flow.

A high level of gene flow into willow populations was also reported by Kikuchi et al. (2011), Trybush et al. (2012) and Perdereau et al. (2014). Due to seed dispersal by convective air, the parent trees of saplings at higher elevations may also be located in lowlands in our study. However, Imbert and Lefèvre (2003), Petit et al. (2005), Hoshikawa et al. (2012) and Perdereau et al. (2014) agree on the comparatively minor role of gene flow through seed compared to gene flow via pollen. Perdereau et al. (2014) found the gene flow rate for *S. caprea* by pollen to be seven times higher than by seed. In their study, seed dispersal comprised about 13 % of the total gene flow, but the authors observed gene flow via goat willow seeds and pollen of more than 200 km, which corresponds to an unimpeded gene flow. Gene flow via seed is characterized by non-random spatial patterns due to the influence exerted by wind or water (Imbert and Lefèvre 2003; Wagner et al. 2004). However, gene flow can also result in a rather random pattern in case of strong microsite influence and seedling mortality (Cortés et al. 2014; Nathan and Muller-Landau 2000). Thus, the results of our study reflected a rather random pattern.

The high allele diversity and the richness of the studied sapling population argue against the genetic isolation of the small parent population within the study site, although there is a general lack of *S. caprea* trees in the Thuringian Forest Mountains overall. The high genetic variation within the population, influenced by external gene flow, cross breeding and admixture (Ojango et al. 2011), complies with the findings of Imbert and Lefèvre (2003), Palmé et al. (2003), Kikuchi et al. (2011), Trybush et al. (2012) and Perdereau et al. (2014), all of whom found only slight or no genetic differences within European willow populations. The findings of these studies therefore support the results of our parentage analysis and the assumption of external gene flow.

Results from parentage analysis, seed trapping and geostatistical modelling all suggested that *S. caprea* seeds may disperse as far as 800 m from their seed source. However, as the search zone for potential parent trees was limited to 500 m around the open areas at our study sites, we were unable to detect dispersal distances longer exceeding 800 m due to the limitations of our study design, even though goat willow seeds are likely dispersed over considerably longer distances.

2.6 Conclusions for silvicultural practice

Often, a rapid natural or manual reforestation after disturbance is required by law. In the context of climate change, the reforestation of disturbed sites becomes even more important. Up until now, too little was known about the seed dispersal distances of *S. caprea* in order to be able to estimate the inexpensive natural regeneration capacity of the species.

Assuming comparable wind patterns, the measured *S. caprea* seed numbers of 1-4 n per trap at distances far from seed trees ($\geq$ 350 m) demonstrate in a central European context that (1) the azimuth direction has no significant influence on the seed dispersal direction of individual trees, (2) the relative position (direction) of seed trees is not important for the number of deposited seeds on disturbed sites, and (3) the number of seed trees has no meaningful influence on seed numbers at a distance of more than 50 m from seed sources, which refute hypothesis 4. Parentage analyses confirmed the (4) important role of an external gene flow for the regeneration observed at study site D; this parent population comprised more goat willow trees than were locally mapped within the 500 m forested search zone. The study of seed dispersal and the genetic analyses revealed a previously underestimated dispersal potential of *S. caprea*.

In the context of the reforestation of disturbed sites by *S. caprea*, the measured seed numbers should be sufficient to establish a natural regeneration layer (independent of mast and non-mast years), if appropriate consideration is given to browsing (Chantal and Granström 2007) and herb competition pressure (= sapling mortality), a sufficient number of microsites ('safe sites') exists for successful germination (Harper 1977), and optimal site and climatic conditions prevail during the germination and establishment phase (see Junttila 1976; Densmore and Zasada 1983; Sacchi and Price 1992; Young and Clements 2003).

Ecological research on germination and sapling development has shown that goat willows germinate immediately after being deposited on bare mineral soil without a humus and litter layer, with good water availability and open-area radiation levels of more than 20 % (Gage and Cooper 2005; Mihók et al. 2005). If conditions are unfavourable, seeds either fail to germinate or seedling mortality may reach 100 % (Densmore and Zasada 1983; Sacchi and Price 1992; Seiwa et al. 2008). Therefore, seed dispersal distance (i.e., seed availability) is not the factor limiting the natural regeneration of goat willow on regeneration sites; rather, unfavourable germination conditions and sapling mortality may be restricting factors.

Any 'spatial optimization' with respect to the position of parent trees by means of forest management is unnecessary due to the omnipresence of willow seeds at the study sites. However, silvicultural practice could integrate measures for conservation, vitalization and propagation of willow seed trees at or near storm-exposed sites in spruce forests at higher altitudes, in order to improve the self-regulation potential of these forests and the natural regeneration of future disturbed areas.

2.7 References

Argus GW (2006) Guide to *Salix* (willow) in the Canadian Maritime Provinces (New Bruns-

wick, Nova Scotia, and Prince Edward Island). E-Publishing, Ontario. http://accs.uaa.alask a.edu/files /botan y/publi catio ns/2006/Guide Salix Canad ianAt lanti cMari time.pdf. Accessed 10 Feb 2018

Barker JHA, Pahlich A, Trybush S, Edwards KJ, Karp A (2003) Microsatellite markers for diverse Salix species. Mol Ecol Notes 3:4–6. https://doi.org/10.1046/j.1471-8286.2003.00332 .x

Barnes BV, Zak DR, Denton SR, Spurr SH (1998) Forest Ecology, 4th edn. Wiley, New York

Barsoum N (2002) Relative contributions of sexual and asexual regeneration strategies in *Populus nigra* and *Salix alba* during the first years of establishment on a braided gravel bed river. Evol Ecol 15:255–279. https://doi.org/10.1023/A:10160 28730 129

Bastide JP, van Vredenburch CLH (1970) The influence of weather conditions on the seed production of some forest trees in the Netherlands. Acta Bot Neerlandica 45:461–490

Baum C, Leinweber P, Weih M, Lamersdorf N, Dimitriou I (2009) Effects of short rotation coppice with willows and poplar on soil ecology. Agric For Res 59:183–196. https://doi.org/10.1007/s1110 4-009-9928-x

Berger JO (1985) Statistical decision theory and Bayesian analysis, 2nd edn. Springer, New York. https://doi.org/10.1007/978-1-4757-4286-2

Bjorkbom JC (1971) Production and germination of paper birch seed and its dispersal into a forest opening. Northeastern Forest Experiment Station, Forest Service Research Paper NE-209, U.S. Department of Agriculture, Upper Darby

Boland JM (2014) Secondary dispersal of willow seeds: sailing on water into safe sites. Madroño 61(4):388–398. https://doi.org/10.3120/0024-9637-61.4.388

Brouwer W, Stählin A (1975) Handbuch der Samenkunde für Landwirtschaft. DLG-Verlags-GmbH, Frankfurt (Main), Gartenbau und Forstwirtschaft

Brown PE (2015) Model-based geostatistics the easy way. J Stat Softw 63(12):1–24. https://doi.org/10.18637 /jss.v063.i12

Bürger M (2003) Bodennahe Windverhältnisse und windrelevante Reliefstrukturen. In: Leibnitz-Institut für Länderkunde (ed) Nationalatlas Bundesrepublik Deutschland. Spektrum Akademischer Verlag, Heidelberg, pp 52–55

Burse K-D, Schramm H-J, Geiling S, Meinhardt H, Schölch M (1997) Die forstlichen Wuchsbezirke Thüringens - Kurzbeschreibung. Mitteilungen der Landesanstalt für Wald und Forstwirtschaft, Gotha

Bushart M, Suck R (2008) Potenzielle natürliche Vegetation Thüringens. Schriftenr. der Thüringer Landesanstalt für Umwelt und Geologie, Jena

Chantal M, Granström A (2007) Aggregations of dead wood after wildfire act as browsing refugia for seedlings of *Populus tremula* and *Salix caprea*. For Ecol Manag 250:3–8. https://doi.org/10.1016/j.forec o.2007.03.035

Chilès JP, Delfiner P (1999) Geostatistics: Modeling spatial uncertainty. Wiley, New York

Chmelar J, Meusel W (1986) Die Weiden Europas, 3rd edn. A. Ziemsen Verlag, Wittenberg

Clark JS, Silman M, Kern R, Macklin E, HilleRisLambers J (1999) Seed dispersal near and far: patterns across temperate and tropical forests. Ecology 80:1475–1494. https://doi.org/10.1890/0012-9658(1999)080%5b147 5:SDNAF P%5d2.0.CO;2

Cortés AJ, Waeber S, Lexer C, Sedlacek J, Wheeler JA, van Kleunen M, Bossdorf O, Hoch G, Rixen C, Wipf S, Karrenberg S (2014) Small-scale patterns in snowmelt timing affect gene flow and the distribution of genetic diversity in the alpine dwarf shrub *Salix herbacea*. Heredity 113:233–239. https://doi.org/10.1038/hdy.2014.19

Densmore R, Zasada J (1983) Seed dispersal and dormancy patterns in northern willows: ecological and evolutionary significance. Can J Bot 61:3207–3216. https://doi.org/10.1139/b83-358

Dickmann DI, Kuzovkina J (2014) Poplars and willows of the world, with emphasis on silviculturally important species. In: Isebrands JG, Richardson J (eds) Poplars and Willows-Trees for society and the environment. FAO, Rome, pp 8–91

Dörken VM (2011) *Salix caprea*-Sal-Weide, Palm-Weide (*Salicaceae*). Jahrb Boch Bot Ver 2:253–257

Dötterl S, Glück U, Jürgens A, Woodring J, Aas G (2014) Floral reward, advertisement and attractiveness to honey bees in dioecious *Salix caprea*. PLoS One 9:e93421. https://doi.org/10.1371/journ al.pone.00934 21

Fink AH, Brücher T, Ermert V, Krüger A, Pinto JG (2009) The European storm Kyrill in January 2007: synoptic evolution, meteorological impacts and some considerations with respect to climate change. Nat Hazards Earth Syst Sci 9:405–423. https://doi.org/10.5194/nhess -9-405-2009

Fischer H, Huth F, Hagemann U, Wagner S (2016) Developing restoration strategies for temperate forests using natural regeneration processes. In: Stanturf JA (ed) Restoration of Boreal and Temperate Forests. CRC Press, Boca Raton, pp 103–164

Frischbier N, Profft I, Hagemann U (2014) Potential impacts of climate change on forest habitats in the Biosphere Reserve Vessertal-Thuringian Forest in Germany. In: Rannow S, Neubert M (eds) Managing protected areas in Central and Eastern Europe under climate change, Advances in Global Change Research. Springer, Dordrecht, pp 243–257. https://doi.org/10.1007/978-94-007-7960-0_16

Füssel U (2007) Floral scent in *Salix* L. and the role of olfactory and visual cues for pollinator attraction of *Salix caprea* L. book, University of Bayreuth

Gage EA, Cooper DJ (2005) Patterns of willow seed dispersal, seed entrapment, and seedling establishment in a heavily browsed montane riparian ecosystem. Can J Bot 83:678–687. https://doi.org/10.1139/B05-042

Gauer J, Aldinger E (2005) Waldökologische Naturräume Deutschlands: Forstliche Wuchsgebiete und Wuchsbezirke-mit Karte 1:1.000.000, Mitteilungen des Vereins für Forstliche Standortskunde und Forstpflanzenzüchtung, Stuttgart

Greene DF, Johnson EA (1996) Wind dispersal of seeds from a forest into a clearing. Ecology 77:595–609. https://doi.org/10.2307/22656 33

Hanley SJ, Mallott MD, Karp A (2006) Alignment of a *Salix* linkage map to the *Populus* genomic sequence reveals macrosynteny between willow and poplar genomes. Tree Genet. Genomes 3:35–48. https://doi.org/10.1007/s1129 5-006-0049-x

Harper JL (1977) Population biology of plants. Academic Press, New York

Herrera CM, Jordano P, Guitián J, Traveset A (1998) Annual variability in seed production by woody plants and the masting concept: reassessment of principles and relationship to pollination and seed dispersal. Am Nat 152:576–594. https://doi.org/10.1086/28619 1

Horvat-Marolt S (1974) Pionirski gozd in iva kot Pionirska drevesna Vrsta II. del - Konkurenčna moč ive (*S. caprea*) kot pionirja v pedosferi (Pioneer forest and sallow (*Salix caprea*) as a pioneer tree species II. Competitive strength of the sallow as a pioneer in the pedosphere). Zb Gozdarstva Lesar 12:5–40

Hoshikawa T, Nagamitsu T, Tomaru N (2012) Effects of pollen availability on pollen immigration and pollen donor diversity in riparian dioecious trees (*Salix arbutifolia*). Botany 90:481–489. https://doi.org/10.1139/B2012 -021

Houle G (1998) Seed dispersal and seedling recruitment of *Betula alleghaniensis*: spatial inconsistency in time. Ecology 79:807–818. https://doi.org/10.1890/0012-9658(1998)079%5b0807:SDASR O%5d2.0.CO;2

Hughes JW, Fahey TJ (1988) Seed dispersal and colonization in a disturbed northern hardwood forest. Bull Torrey Bot Club 115(2):89–99

Huth F (2009) Untersuchungen zur Verjüngungsökologie der Sand-Birke (*Betula pendula* Roth). University of TU-Dresden, PhD-book

Imbert E, Lefèvre F (2003) Dispersal and gene flow of *Populus nigra* (*Salicaceae*) along a dynamic river system. J Ecol 91:447–456. https://doi.org/10.1046/j.1365-2745.2003.00772 .x

Junttila O (1976) Seed germination and viability in five *Salix* species. Astarte 9:19–24

Karlsson M (2001) Natural regeneration of broadleaved tree species in southern Sweden: effects of silvicultural treatments and seed dispersal from surrounding stands. book, Swedish University of Agricultural science

Karrenberg S, Suter M (2003) Phenotypic trade-offs in the sexual reproduction of *Salicaceae* from flood plains. Am J Bot 90:749–754. https://doi.org/10.3732/ajb.90.5.749

Karrenberg S, Kollmann J, Edwards PJ (2002) Pollen vectors and inflorescence morphology in four species of *Salix*. Plant Syst Evol 235:181–188. https://doi.org/10.1007/s0060 6-002-0231-z

Kay QON (1985) Nectar from willow catkins as a food source for Blue Tits. Bird Study 32:40–44. https://doi.org/10.1080/0006365850 94768 53

Kelly D (1994) The evolutionary ecology of mast seeding. Tree 9:465–470

Kelly D, Sork VL (2002) Mast seeding in perennial plants: why, how, where? Annu Rev Ecol Syst 33:427–447. https://doi.org/10.1146/annur ev.ecols ys.33.02060 2.09543 3

Kikuchi S, Suzuki W, Sashimura N (2011) Gene flow in an endangered willow *Salix hukaoana* (*Salicaceae*) in natural and fragmented riparian landscapes. Conserv Genet 12:79–89. https://doi.org/10.1007/s1059 2-009-9992-z

Kohlermann L (1950) Untersuchungen über die Windverbreitung der Früchte und Samen mitteleuropäischer Waldbäume. Forstwiss Cent 69:606–624. https://doi.org/10.1007/BF018 15738

Kollmann J, Goetze D (1998) Notes on seed traps in terrestrial plant communities. Flora 193:31–40. https://doi.org/10.1016/S0367-2530(17)30813 -7

Kolodziej A, Frühauf M (2008) Phänologische Veränderungen wild wachsender Pflanzen in Sachsen-Anhalt 1962-2005. Hercynia NF 41:23–37

Küßner R (1997) Sukzessionale Prozesse in Fichtenbeständen (*Picea abies*) des Osterzgebirges: Möglichkeiten ihrer waldbaulichen Beeinflussung und ihrer Bedeutung für einen ökologisch begründeten Waldumbau. Forstwiss Cent 116:359–369

Kuzovkina YA, Quigley MF (2005) Willows beyond wetlands: uses of *Salix* L. species for environmental projects. Water Air Soil Pollut 162:183–204. https://doi.org/10.1007/s11270-005-6272-5

Lautenschlager D (1994) Die Weiden von Mittel- und Nordeuropa, überarbeitete edn. Birkhäuser Verlag, Basel

Leder B (1992) Weichlaubhölzer: Verjüngungsökologie, Jugendwachstum und Bedeutung in Jungbeständen der Hauptbaumarten Buchen und Eiche, Schriftenreihe der Landesanstalt für Forstwissenschaft Nordrhein-Westfalen

Matlack GR (1989) Dispersal of seed across snow in *Betula lenta*, a gap-colonizing tree species. J Ecol 77:853–869. https://doi.org/10.2307/22609 90

McVean DN (1953) *Alnus glutinosa* (L.) Gaertn. J Ecol 41:447–466. https://doi.org/10.2307/22570 70

McVean DN (1956) Ecology of *Alnus glutinosa* (L.) Gaertn: VI. Post-Glacial History. J Ecol 44:331–333. https://doi.org/10.2307/22568 25

Mihók B, Gálhidy L, Kelemen K, Standovár T (2005) Study of gap phase regeneration in a managed beech forest: relations between tree regeneration and light, substrate features and cover of ground vegetation. Acta Silv Lignaria Hung 1:25–38

Moon K, Duff TJ, Tolhurst KG (2013) Characterising forest wind profiles for utilisation in fire spread models. In: Piantadosi J, Anderssen RS, Boland J (eds) MODSIM2013, Modelling and simulation society of Australia and New Zealand. Presented at the 20th International Congress on Modelling and Simulation, Adelaide, Australia, pp 214–220

Mosseler A, Papadopol CS (1989) Seasonal isolation as a reproductive barrier among sympatric *Salix* species. Can J Bot 67:2563–2570. https://doi.org/10.1139/b89-331

Nathan R, Muller-Landau HC (2000) Spatial patterns of seed dispersal, their determinants and consequences for recruitment. Tree 15:278–285. https://doi.org/10.1016/S0169 -5347(00)01874 -7

Nei M (1972) Genetic distance between populations. Am Nat 106:283–392

Nei M (1978) Estimation of average heterozygosity and genetic distance from a small number of individuals. Genetics 89:583–590

Niiyama K (1990) The role of seed dispersal and seedling traits in colonization and coexistence of *Salix* species in a seasonally flooded habitat. Ecol Res 5:317–331. https://doi.org/10.1007/BF023 47007

Ojango JM, Mpofu N, Marshall K, Andersson-Eklun L (2011) Quantitative methods to improve the understanding and utilisation of animal genetic resources. In: Ojango JM, Malmfors B, Okeyo AM (eds) Animal genetics training resource, Version 3. International Livestock Research Institute, Nairobi, Kenya, and Swedish University of Agricultural Sciences, Uppsala, Sweden, pp 1–39

Okubo A, Levin SA (1989) A theoretical framework for data analysis of wind dispersal of seeds and pollen. Ecology 70:329–338. https://doi.org/10.2307/19375 37

Palmé AE, Semerikov V, Lascoux M (2003) Absence of geographical structure of chloroplast DNA variation in sallow, *Salix caprea* L. Heredity 91:465–474. https://doi.org/10.1038/sj.hdy.68003 07

Peakall R, Smouse PE (2012) GenAlEx 6.5: genetic analysis in Excel. Population genetic software for teaching and research-an update. Bioinformatics 28:2537–2539. https://doi.org/10.1093/bioin forma tics/bts46 0

Perala DA, Alm AA (1990) Reproductive ecology of birch: a review. For. Ecol. Manag. 32:1–38. https://doi.org/10.1016/0378-1127(90)90104 -J

Perdereau AC, Kelleher CT, Douglas GC, Hodkinson TR (2014) High levels of gene flow and genetic diversity in Irish populations of *Salix caprea* L. inferred from chloroplast and nuclear SSR markers. Plant Biol 14:1–12. https://doi.org/10.1186/s12870-014-0202-x

Petit RJ, Duminil J, Fineschi S, Hampe A, Salvini D, Vendramin GG (2005) Comparative organization of chloroplast, mitochondrial and nuclear diversity in plant populations. Mol Ecol 14:689–701. https://doi.org/10.1111/j.1365-294X.2004.02410 .x

Regvar M, Likar M, Piltaver A, Kugonic N, Smith JE (2010) Fungal community structure under goat willows (*Salix caprea* L.) growing at metal polluted site: the potential of screening in a model phytostabilisation study. Plant Soil 330:345–356. https://doi.org/10.1007/s1110 4-009-0207-7

Ribbens E, Silander JA Jr, Pacala SW (1994) Seedling recruitment in forest: calibrating models to predict patterns of tree seedling dispersion. Ecology 75:1794–1806. https://doi.org/10.2307/1939638

Richardson J, Isebrands JG, Ball JB (2014) Ecology and physiology of poplars and willows. In: Richardson J (ed) Isebrands JG. Trees for Society and the Environment, Poplars and Willows, pp 92–123

Robert CP (2007) The Bayesian choice: from decision-theoretic foundations to computational implementation, 2nd edn. Springer, New York. https://doi.org/10.1007/0-387-71599 -1

Rue H, Martino S, Chopin N (2009) Approximate Bayesian inference for latent Gaussian models by using integrated nested Laplace approximations. J. R. Statist. Soc. B 71:319–392. https://doi.org/10.1111/j.1467-9868.2008.00700 .x

Ryvarden L (1971) Studies in seed dispersal I. Trapping of diaspores in the alpine zone at Finse, Norway. Nor J Bot 18:215–226

Sacchi CF, Price PW (1992) The relative roles of abiotic and biotic factors in seedling demography of arroyo willow (*Salix lasiolepis: Salicaceae*). Am J Bot 79:395–405. https://doi.org/10.1002/j.1537-2197.1992.tb145 66.x

Sarvas R (1952) On the flowering of birch and the quality of seed crop. Commun Inst For Fenn 40:1–35

Scheffler A, Frühauf M (2011) Veränderungen der Pflanzenphänologie in unterschiedlichen Naturräumen Sachsen-Anhalts unter Berücksichtigung ihrer wesentlichen Einflussfaktoren. Hercynia NF 44:169–189

Schirmer R (2006) *Salix alba* Linné. In: Schütt P, Weisberger H, Schuck HJ, Lang U, Stimm B (eds) Enzyklopädie Der Laubbäume. Nikol Verlagsgesellschaft mbH & Co KG, Hamburg, pp 535–550

Schmiedel D, Huth F, Wagner S (2013) Using data from seed-dispersal modelling to manage invasive tree species: the example of *Fraxinus pennsylvanica* Marshall in Europe. Environ Manag 52:851–860. https://doi.org/10.1007/s0026 7-013-0135-4

Schütt P (2006) *Salix caprea* LINNÉ, 1753. In: Schütt P, Weisberger H, Schuck HJ, Lang U, Stimm B (eds) Enzyklopädie Der Laubbäume. Nikol Verlagsgesellschaft mbH & Co., KG, Hamburg, pp 551–558

Seiwa K, Tozawa M, Ueno N, Kimura M, Yamasaki M, Maruyama K (2008) Roles of cottony hairs in directed seed dispersal in riparian willows. Plant Ecol 198:27–35. https://doi.org/10.1007/s1125 8-007-9382-x

Simpson D, Rue H, Riebler A, Martins TG, Sørbye SH (2017) Penalising model component complexity: a principled, practical approach to constructing priors. Stat Sci 32:1–28. https://doi.org/10.1214/16-STS57 6

Skarpaas O, Auhl R, Shea K (2006) Environmental variability and the initiation of dispersal: turbulence strongly increases seed release. Proc Biol Sci 273:751–756. https://doi.org/10.1098/rspb.2005.3366

Skvortsov AK (1999) Willows of Russia and adjacent countries-taxonomical and geographical revision, University of Joensuu, Faculty of mathematics and natural sciences report series. Finland

Stoyan D, Wagner S (2001) Estimating the fruit dispersion of anemochorous forest trees. Ecol Model 145:35–47. https://doi.org/10.1016/S0304 -3800(01)00385 -4

Trybush SO, Jahodová Š, Čížková L, Karp A, Hanley SJ (2012) High levels of genetic diversity in *Salix viminalis* of the Czech Republic as revealed by microsatellite markers. Bioenergy Res. 5:969–977. https://doi.org/10.1007/s1215 5-012-9212-4

van Splunder I, Coops H, Voesenek LACJ (1995) Establishment of alluvial forest species in floodplains: the role of dispersal timing, germination characteristics and water level fluctuations. Acta Bot Neerlandica 44:269–278. https://doi.org/10.1111/j.1438-8677.1995.tb007 85.x

Vroege PW, Stelleman P (1990) Insect and wind pollination in *Salix repens* L. and *Salix caprea* L. Isr J Bot 39:125–132. https://doi.org/10.1080/00212 13X.1990.10677 137

Waesch G (2003) Montane Graslandvegetationen des Thüringer Waldes: Aktueller Zustand, historische Analyse und Entwicklungsmöglichkeiten, 1st edn. Cuvillier Verlag, Göttingen

Wagner S (1997) Ein Modell zur Fruchtausbreitung der Esche (*Fraxinus excelsior* L.) unter Berücksichtigung von Richtungseffekten. Allg. Forst- Jagdztg. 168:149–155

Wagner S, Wälder K, Ribbens E, Zeibig A (2004) Directionality in fruit dispersal models for anemochorous forest trees. Ecol Model 179:487–498. https://doi.org/10.1016/j.ecolm odel.2004.02.020

Werner P (1975) A seed trap for determining patterns of seed deposition in terrestrial plants. Can J Bot 53:810–813. https://doi.org/10.1139/b75-097

Worrell R (1995) European aspen (*Populus tremula* L.): a review with particular reference to Scotland I. Distribution, ecology and genetic variation. Forestry 68:93–105. https://doi.org/10.1093/fores try/68.2.93

Young JA, Clements CD (2003) Seed germination of willow species from a desert riparian ecosystem. J Range Manag 56:496–500. https://doi.org/10.2307/40038 42

Zar JH (2010) Biostatistical analysis, 5th edn. Prentice Hall, Upper Saddle River

Zasada JC, Densmore R (1979) A trap to measure *Populus* and *Salix* seedfall. Can Field-Nat 93:77–79

Zerbe S (2009) Renaturierung von Waldökosystemen. In: Zerbe S, Wiegleb G (eds) Renaturierung von Ökosystemen in Mitteleuropa. Spektrum Akademischer Verlag, Heidelberg, pp 153–182

Ziello C, Estrella N, Kostova M, Koch E, Menzel A (2009) Influence of altitude on phenology of selected plant species in the Alpine region (1971–2000). Clim Res 39:227–234. https://doi.org/10.3354/cr008 22

Żywiec M, Ledwoń M (2008) Spatial and temporal patterns of rowan (*Sorbus aucuparia* L.) regeneration in West Carpathian subalpine spruce forest. Plant Ecol 194:283–291. https://doi.org/10.1007/s1125 8-007-9291-z

Żywiec M, Holeksa J, Wesołowska M, Szewczyk J, Zwijacz-KozicaT, Kapusta P, Bruun HH (2013) *Sorbus aucuparia* regeneration in a coarse-grained spruce forest-a landscape scale. J Veg Sci 24:735–743. https://doi.org/10.1111/j.1654-1103.2012.01493 .x

Chapter 3

Restrictions on natural regeneration of storm-felled spruce sites by silver birch (*Betula pendula* Roth) through limitations in fructification and seed dispersal

Katharina Tiebel[1], Franka Huth[1], Nico Frischbier[2], Sven Wagner[1] (under review)

European Journal of Forest Research

[1]TU Dresden, Institute of Silviculture and Forest Protection, Chair of Silviculture, Pienner Str. 8, 01737 Tharandt, Germany

[2]ThüringenForst, Forestry Research and Competence Center, Jägerstraße 1, 99867 Gotha, Germany

3.2 Introduction

As an anemochorously dispersed pioneer tree species with a wide natural range over Eurasia, silver birch (*Betula pendula* Roth) has a high ecological value within temperate and boreal forest ecosystems (Atkinson 1992; Hynynen et al. 2010). Silver birches enhance soil nutrition and soil stability, provide watershed protection, act as structural elements with a long-term stabilizing effect, and provide habitats and food for many organisms (see Patterson 1993; Humphrey et al. 1998; Ferris and Humphrey 1999; Priha 1999; Beck et al. 2016). In some European countries, like England, Sweden, Finland and Latvia, silver birch is the most im-portant broadleaved tree species for timber production, plywood or veneer production (Cam-eron 1996; Luostarinen and Verkasalo 2000; Hynynen et al. 2010). Within their natural geographical range in Europe, birch species were often considered as a forest weed and, therefore, rigorously thinned out of forest stands during the last century (Röhrig and Gussone 1990; Koski and Rousi 2005). However, silver birch recently received renewed interest with respect to forest management at higher altitudes. The importance of silver birch in forest management has recently been increasing because of (a) the species' ability to promptly and extensively recolonize disturbed sites due to its high annual seed production and its fast juvenile growth, even in open areas with extreme climatic conditions (Perala and Alm 1990; Atkinson 1992; Zerbe 2001; Hynynen et al. 2010), and (b) the heightened risk of catastrophic events in central European spruce forests. Pioneer forests composed of birches are able to quickly close water and nutrient cycles and thus soon create a forest climate appropriate for the establishment of climax tree species (Zerbe 2009).

Therefore, empirical information about the seed production, seed dispersal distances, and deposited seed numbers of silver birch is required to establish 'precautionary' forest management systems that anticipate the high risk of catastrophic events, particularly in mountain spruce forests, and to ensure successful birch regeneration on disturbed sites. The seed production of a mature single silver birch tree can range between 30,000 and 10 million seeds per year (Arnborg 1948 cited in Perala and Alm 1990; Popadyuk et al. 1995; Huth 2009). The small winged nuts (1.5-2.0 mm) are mainly dispersed by wind between June and November (Brouwer and Stählin 1975; Huth 2009).

Above all, seed dispersal is an important driver for species movement, site colonization and the restoration of treeless or disturbed areas (Skarpaas et al. 2006). Huth (2009) reported mean seed dispersal distances of 37 to 90 m for admixed silver birch trees within closed Norway spruce forests. Different studies determined the highest birch seed densities within distances of 25 to 50 m around the source trees (Sarvas 1948; Fries 1984). The phenomenon of secondary seed dispersal by wind after a transitory deposition is important, in particular for the transport of birch seeds in large restoration areas (Matlack 1989; Bakker et al. 1996). For *B. lenta*, the secondary seed dispersal distance across snow was three times longer than the measured primary seed dispersal (Matlack 1989), but secondary seed dispersal distance reached only 15 m when the snow was melting (Greene and Johnson 1997). However, most studies on aspects of birch seed distribution have been conducted within closed forest stands (e.g., Skoglund and Verwijst 1989; Houle and Payette 1990; Graber and Leak 1992; Leder 1992; Houle 1998; Wagner et al. 2004; Huth 2009). Only few studies focused on the seed dispersal of birch in open areas or large gaps (Bjorkbom 1971; Hughes and Fahey 1988; Greene and Johnson 1996; Karlsson 2001), although knowledge about the seed dispersal distance from surrounding forest stands and seed distribution limits is necessary key information to develop recolonization and restoration management strategies (see Zhao et al. 2016; Holmström et al. 2017).

As shown in different wind tunnel experiments, experimental results under controlled conditions cannot easily be transferred to real field conditions (Augspurger and Franson 1987; Johnson and West 1988, cited in Bakker et al. 1996; Kadereit and Leins 1988; van Dorp et al. 1996; Greene and Johnson 1997). This lack of transferability is caused by highly complex and variable environmental factors related to field conditions (e.g., wind conditions, site surface relief and ground vegetation cover or seed characteristics) (Fenner 1985; Okubo and Levin 1989; Skarpaas et al. 2006). Therefore, reliable information about the temporal and spatial patterns of seed rain in open areas is needed to assess the natural regeneration potential of *B. pendula* seedlings. This applies particularly to the large windthrown forest areas in central Europe, which were created by the storm events of the last decades (Gregow et al. 2017) and will probably become more and more frequent with progressing climate change (Mölter et al. 2016).

In this case study, we observed the seed dispersal of *B. pendula* in 2015 and 2016 at two windthrown forest sites in Thuringia, Germany. The aim of the study was to investigate the amounts, densities and spatial distribution of silver birch seeds in storm-felled, treeless areas at high altitudes (715-775 m a.s.l.) originating from adjacent closed mountain forests. Initially, the temporal and spatial patterns of silver birch seeds were empirically recorded, with subsequent calculation of mean dispersal distances (MDD) by means of inverse modelling. We assume that models showed directionality (anisotropy) for birch seed dispersal. Furthermore, the influence of seed crop, relief inclination and seed tree numbers around the studied storm-felled sites as well as the position of the seed trees (valley, slope or plateau) were included in the analyses. Finally, we used simulations to spatially optimize the positioning of the seed trees in relation to the studied sites with regard to optimal seed distribution in the open areas.

3.3 Materials and methods

3.3.1 Study area

The study area is located at high elevations and along the ridges of the Thuringian Forest, a mountain range in the federal state of Thuringia, Germany (50°40'N and 10°45'E). The area is situated between 400-982 m above sea level (a.s.l.), with a prevailing south-westerly exposition. The area is characterized by many slopes and an almost total absence of plateaus (Burse et al. 1997; Waesch 2003; Gauer and Aldinger 2005). The mean annual precipitation ranges from 800 mm in the south-west to 1,200 mm along the ridges and falls to a level of 700 mm in the north-east (Burse et al. 1997; Gauer and Aldinger 2005; Bushart and Suck 2008). The annual average temperature in the region varies between 4-6 °C (Burse et al. 1997; Bushart and Suck 2008). The area is influenced by an Atlantic, moderately cool and moist central mountain climate (Burse et al. 1997; Gauer and Aldinger 2005). The prevailing winds are from the south-west, with a secondary wind maximum originating from the north-east. The average annual wind speed in the study area is 3.5-4.5 m s^{-1} (Bürger 2003). The averaged meteorological data (based on half-hourly values) for the seed trapping periods of the presented study (2015 and 2016) are listed in Table 3.1. While no extreme events in wind speed were observed, the wind direction showed a high variability during the study periods.

Table 3.1 Aggregated meteorological data (based on half-hourly values; climate station 'Grosser Eisenberg'; 50° 37' 24″ N and 10° 46' 59″ O) of the four-month seed-trapping periods in 2015 and 2016 in the study area. Please note that the data of July and November covers only studied days and not the entire month (NA - data not available due to measurement failures, SD - standard deviation).

	2015					2016				
	from mid-Jul	Aug	Sep	Oct	until early-Nov	from mid-Jul	Aug	Sep	Oct	until early-Nov
Wind spead (m/s)										
Minimum	1.0	0	0	0	0.7	0	0.6	0	0	1.2
Maximum	7.2	8.3	9.9	4.6	4.4	4.7	5.2	6.4	6.5	3.5
Mean	3.6	2.9	1.8	2.0	2.4	1.8	2.2	2.1	2.3	2.5
SD	1.42	1.22	0.92	0.88	0.78	0.75	0.87	1.06	1.11	0.62
Wind direction (°)										
Mean	228	159	204	180	237	184	218	178	176	273
Median	253	181	235	214	233	226	250	206	222	283
SD	86.7	98.5	122.2	100.8	24.7	105.2	89.6	102.2	106.2	53.3
Mean temperature (°C)	11.1	18.2	9.8	6.0	9.9	17.7	14.9	14.4	5.1	3.8
Precipitation (mm/month)	0.6	65.2	74.3	NA	0.4	70.6	69.6	74.8	127.9	4.5

The dominant soil types of the forest sites are low-base cambisols with low to medium nutrient contents (Gauer and Aldinger 2005). The regional landscape features a largely contiguous forest system with ~90 % forest cover, some small upland meadows in stream valleys and occasional small raised bogs. The study area is dominated by single-layered, even-aged Norway spruce forests (*Picea abies* (L.) Karst.). Without anthropogenic influence, the potential natural vegetation would be dominated by *Luzulo-Fagetum* and *Asperulo-Fagetum* beech forests (Frischbier et al. 2014).

We selected two study sites (B and E) 6 km apart from each other, located on slopes at higher elevations of the Thuringian Forest (715-775 m a.s.l.). Each site consisted of an open area surrounded by a forested search zone of 200 m (see chapter 3.3.2, p. 63). The open areas were windthrown by the storm 'Kyrill' in January 2007 (Fink et al. 2009). Representative for the region, the stand conditions before the storm were dominated by 68-100 year-old Norway spruce. After the storm, the damaged areas were completely cleared, and no birch seed trees were present in the open areas. The size of the open areas was 4.0 ha and 12.7 ha, respectively (Table 3.2), and no closed regeneration layer had yet established at any of the study sites. The open areas were surrounded by 59-105 year-old Norway spruce forests admixed with a small number of adult isolated *Betula pendula* Roth, *Salix caprea* L. and *Sorbus aucuparia* L. trees.

Table 3.2 Descriptive study site and birch seed tree data (Dbh - diameter at breast high, SD - standard deviation).

| Open area at study sites | B | E |
Relief-induced dispersal	'uphill'	'downhill'
Elevation above sea level [m]	735 - 765	715 - 775
Topography	mountain peak with slopes	slopes
Size of open area [ha]	4.0	12.7
Number of seed traps [n]	54	41
Number of seed trees [n] within the 200 m forested search-zone around the open area [n]	16	83
Minimum distance between seed tree and seed trap [m]	12	74
Average Dbh of seed trees ± SD [cm]	31.1 ± 4.7	20.7 ± 4.4

Located along slopes, the choice of study sites allowed us to separately investigate uphill (site B) and downhill (site E) seed dispersal. The seed trees at site B were located in the valley at approximately 710-730 m a.s.l., and they were equipped with seed traps from the seed sources all the way to the uphill plateau at 760 m a.s.l. (Fig. 3.1). At site E, seed trees were mainly found on a plateau (785-805 m a.s.l.) within a stand consisting of *Salix caprea, Sorbus aucuparia* and *B. pendula*, and seed traps were placed close to the seed sources on the upper slope (775 m a.s.l.) and downhill along the slope to the valley (675 m a.s.l.).

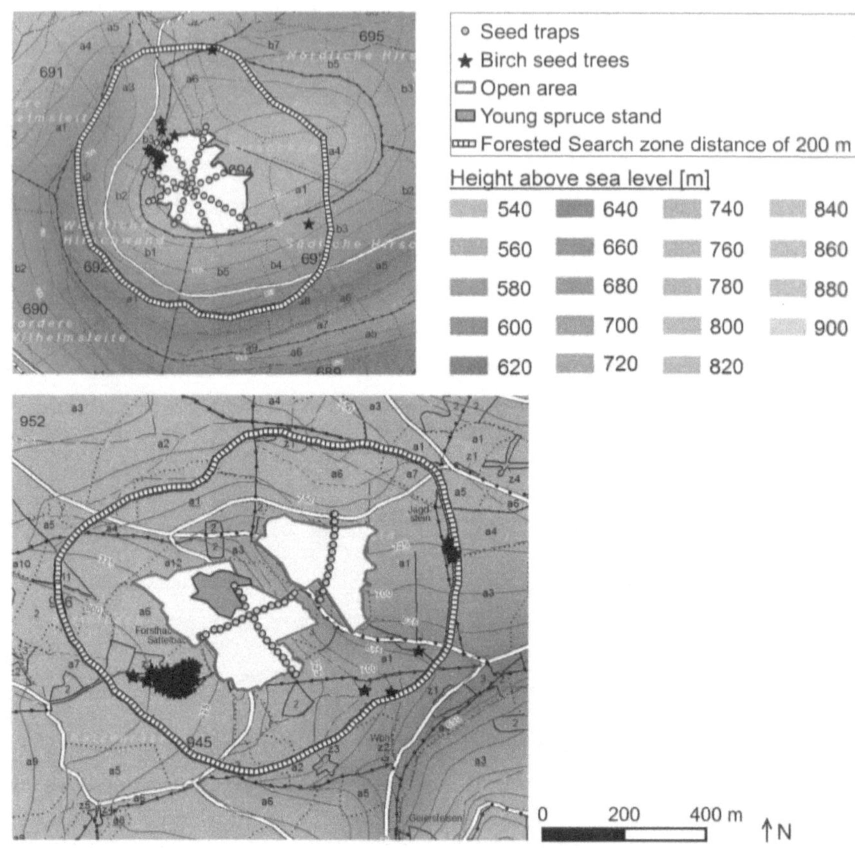

Fig. 3.1 Maps of the experimental study design at sites B (top) and E (bottom).

3.3.2 Experimental design

Within the 200 m forested search zone at each study site, we mapped all *B. pendula* trees that were expected to potentially produce seeds ($\geq$ 12 cm diameter at breast high [dbh]; see Popadyuk et al. 1995; Roloff and Pietzarka 2010) using a blumax Bluetooth GPS-4013 Receiver. For each *B. pendula* tree, we recorded the dbh and observed flowering in both years. The search zone distance of 200 m was chosen as a compromise between feasibility and prior knowledge of suggested effective birch seed dispersal distance in open and forested sites (Sarvas 1948; Karlsson 2001; Huth 2009). Sixteen and 83 seed trees with 24-42 cm and 13-37 cm in dbh were identified at sites B and E, respectively (Table 3.2).

At the study sites, we placed 54 (site B) and 41 seed traps (site E). Due to the vast areal extent of the open areas, seed traps were placed along two crossing line transects (site E) and four crossing line transects (site B), with intervals of 20 m between the traps, rather than along a regular grid (Fig. 3.1 - see also Bjorkbom 1971; Greene and Johnson 1996). The orientation and length of the line transects were not uniform, due to the differences in the size and shape of the two open areas. The line transects extended over the entire open area of each study site and into the surrounding Norway spruce forests. The minimum distances between the seed trees and the nearest seed trap were 12 and 74 m (Table 3.2).

The funnel-shaped seed traps had a diameter of 0.5 m and surface area of 0.196 m². To ensure the functioning of the seed traps despite strong winds, a perforated plastic cup weighted with a stone was placed into each funnel-shaped net. The percolated plastic cups allowed rain water to runoff. The net funnels were fixed onto a bar 1 m above the ground. The traps were emptied periodically every 3 to 4 weeks and the number of seeds per trap was counted. The seed dispersal sampling periods each lasted 4 months from mid-July to early November in 2015 and 2016.

3.3.3 Data analysis

Mean seed densities per m² were calculated for each seed trap across both study sites and years. Differences between seed densities at the two sites and between the two sampling years were analysed using the Mann-Whitney U-test, because the data were not normally distributed (Zar 2010). Significant differences were accepted at a p-value of < 0.05. Furthermore, Quantum GIS (QGIS 2.4.0 Chugiak) was used to create maps of both study sites based on original forest maps and aerial orthophotographs. These maps were used to outline the boundaries of the open areas at each study site, to determine the surrounding 200 m forested search zone, and to mark the positions of seed trees and seed traps (Fig. 3.1).

3.3.4 Seed dispersal model

A phenomenological model (provided as R-script) developed by van Putten et al. (2012) was used to investigate birch seed dispersal, including the effect of wind direction, the probability of seed deposition at certain distances from the seed source and a dbh-related prediction of seeds per tree and per year. The applied model is capable of accounting for the direction of seed dispersal, thus differentiating between isotropic and anisotropic dispersal. 'Isotropic' means that seed densities are equally dispersed in all azimuth directions, whereas 'anisotropic' dispersal accounts for a directional effect (e.g., due to wind) on seed density distributions (Wagner et al. 2004; Wälder et al. 2009). We fitted seed shadows using the isotropic model and the anisotropic *no-shift* elliptic distorted-distance model with the free parameters β,

ψ and γ. The parameter β (coherency) determines the flattening of the elliptic contour lines, ψ (rotation) describes the rotation of the elliptic contour lines around a seed tree along the common axis, and γ (drift) moves the centre of elliptic contour lines into a positive direction along the common axis (van Putten et al. 2012). The most important algorithms were described by van Putten et al. (2012). To model the distance-dependent seed distribution, i.e. the 'kernel' of the model in the Cartesian coordinate system (x, y), we used the negative exponential distribution as a density function ($d(r_{(x,y)})$, Eq. 3.1):

$$d(r_{(x,y)}) = \frac{e^{\left[-\frac{r_{(x,y)}}{\lambda}\right]}}{r_{(x,y)} * 2\pi\lambda} \qquad (3.1)$$

The dispersal distance within the negative exponential function is described by the parameter λ. The value $r_{(x,y)}$ describes the distance between the position of the seed trees and seed traps using the Cartesian coordinates x and y, where seed density is known. Other models, e.g. lognormal, have been tested without improving the results. The fecundity of a seed tree φ was calculated using the following equation (Eq. 3.2):

$$\varphi = e^{\alpha} * dbh^2 \qquad (3.2)$$

with α as a fecundity parameter defining the allometric relationship between the dbh (mm) and the seed production of a tree. Isotropic and anisotropic seed dispersal was modelled separately for each study site (B and E) and year (2015 and 2016). The two study sites were split into the categories 'uphill' (B) and 'downhill' (E) on the basis of the relief-related seed dispersal. Inverse modelling was applied to fit the observed seed densities. The seed number modelled for each seed trap was calculated by summing the seed rain at a specific location relative to all seed trees. The mean dispersal distance (MDD) in the negative exponential kernel equals λ. In the case of isotropic modelling, the parameter equals MDD_{iso}. Spearmann's correlation coefficient including p-value was used to test the relation between observed and predicted seed densities. Additionally, Akaike's information criterion (AIC) was used to check the goodness-of-fit of the statistical models. General references to isotropic and anisotropic inverse modelling can be found in Okubo and Levin (1989), Ribbens et al. (1994), Clark et al. (1999), Skarpaas et al. (2004), Wagner et al. (2004), Soubeyrand et al. (2007), Wälder et al. (2009) and van Putten et al. (2012).

The parametric bootstrap approach, described by Faraway (2006) and Tekle et al. (2016), was used to compare isotropic and anisotropic inverse model fits and to make a decision on the significance level (i.e. which model showed a better fit to the empirical data). The likelihood ratio test (LRT) allows comparing models with different numbers of parameters by means of differences in log likelihood between them. Bootstrap samples (data sets) were generated under a 'null model' (isotropic model) using the estimated parameters. Then 'null' and 'alternative models' (anisotropic model) were then fitted based on these data sets and the likelihood ratio statistic was computed. This procedure was repeated 99 times for each study site and year. The differences in log likelihood between the isotropic and anisotropic models were used to derive an empirical distribution of LRT, where the null-hypothesis was true. The p-value was estimated by comparing the empirical distribution of LRT to the observed values of LRT output (Faraway 2006; Tekle et al. 2016). All computations were performed using the R software version 3.3.2 (package: boot; R Core Team 2014).

3.3.5 Simulations for practical management decisions

To apply our findings of seed dispersal in a practical context and to support silvicultural management decisions in the context of reforesting disturbed sites, two alternative scenarios of seedtree distribution were designed for study sites B and E, based on the area-specific seed dispersal model results with MDDs of 100 and 350 m, respectively. A regular distribution of seed trees on a 100 m grid surrounding the open areas (i.e. 30 trees) was compared with an aggregated seed tree distribution of the same tree numbers. The simulation was done for two conceptual forested sites with a size of 42 ha (700 m x 600 m) in which the two differently sized and shaped open areas from study sites B and E were integrated. All birch seed trees were assumed to have a dbh of 20 cm and to produce 1.5 million seeds as fitted by inverse modelling in 2016.

3.4 Results

3.4.1 Seed production

The densities of the deposited seeds in both open areas and surrounding forests were significantly higher in 2016 than in 2015 (Mann-Whitney U-test: $p < 0.001$). Average seed densities ranged between 93 and 23 n m^{-2} in 2015 and 445 and 86 n m^{-2} in 2016 at sites B and E, respectively (Fig. 3.2). Overall, the recorded birch seed densities in traps were at least four times higher in 2016 than in 2015, with a maximum of 9,557 (site B) and 311 n m^{-2} (site E).

In both years, birch seed numbers tended to be higher at site B, but this difference was not significant (Mann-Whitney U-test: $p > 0.05$).

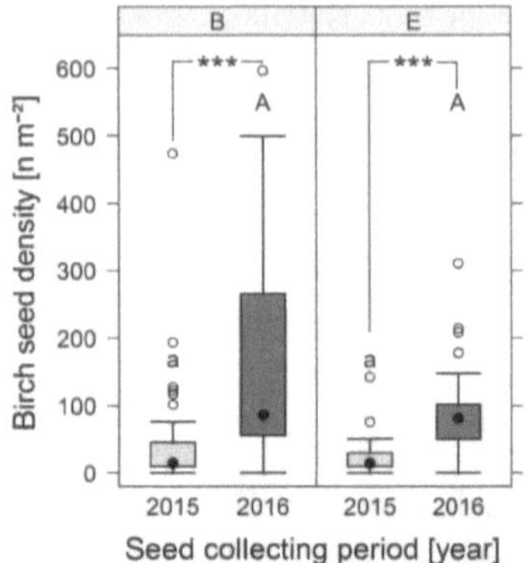

Fig. 3.2 *Betula pendula* seed densities [n m⁻²] collected in 2015 and 2016 at study sites B and E. Lower and upper case letters indicate significant differences between study sites in 2015 and 2016, respectively. Stars mark significant differences between years at each study site (Mann-Whitney U-test: $p < 0.05$). White circles show outliers and black circles inside the boxes are mean values.

The allometric relationships between tree dbh and seed production for isotropic models were very tight. The fecundity levels (i.e. 'α' in Eq. 3.2) for the years 2015 (2.0-2.2) or 2016 (3.6) were relatively similar, indicating a slightly lower fecundity of seed trees in 2015. The expected seed production rate of the isotropic inverse model for a birch seed tree with a mean dbh of 20 cm was approximately 300,000-366,000 (2015) and 1,430,000-1,530,000 seeds per tree (2016) (Table 3.3). Birch seed trees with 13-42 cm in dbh produced 0.14-1.5 million and 0.62-6.5 million seeds per tree in 2015 and 2016, respectively (Fig. 3.3).

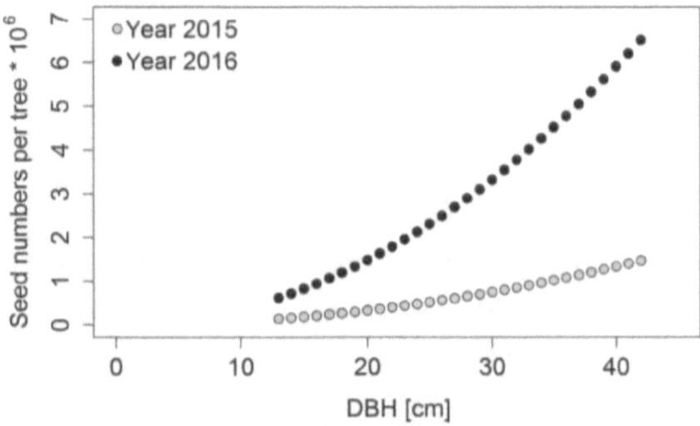

Fig. 3.3 Isotropic inverse model predictions of seed production per seed tree depending on dbh (diameter at breast height) in 2015 and 2016.

3.4.2 Seed dispersal and spatial patterns

The source tree-related pattern of cumulated trapped seed numbers followed a negative exponential function (Fig. 3.4). The highest seed densities at the study sites were found close to the seed sources, e.g., at the northern edge of the open area at site B or close to a stand of *Salix caprea, Sorbus aucuparia* and *B. pendula* within a spruce forest neighbours adjacent to the open area at site E (see Fig. 3.1). Seed densities decreased rapidly with increasing distance from the seed source. At a distance of 100 m from the seed sources, mean seed densities of only 24 and 41 n m^{-2} were observed at site B (uphill dispersal) and E (downhill dispersal) in 2015, respectively, compared to 114 and 181 n m^{-2} in 2016. During the same period, seed densities of only 15-25 n m^{-2} were recorded downhill at a distance of 300 m from the seed sources at site E. In both years, the number of seeds trapped at the same distance from the seed source was slightly higher at site E compared to site B (Fig. 3.4).

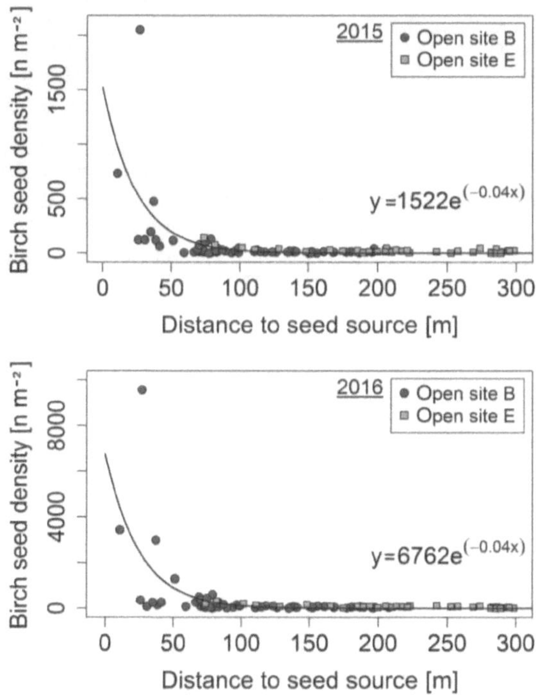

Fig. 3.4 Complete set of measured *Betula pendula* seed densities [n m^{-2}] at study sites B and E in 2015 and 2016 related to the distance from the seed source [m]. Note the different scales of the y-axes.

The probability distribution of the seed dispersal distance of a single tree in the anisotropic models was quite similar to that of the isotropic models, although the predicted seed shadows clearly showed directional variation (Table 3.3 and Fig. 3.5). Values of β, γ, ψ, which characterize the spatial distribution of the anisotropic seed shadows, are listed in Table 3.3. Although the anisotropic models featured higher AIC values (with the exception of study site B in 2016), the bootstrap results indicated that the anisotropic model was over-parameterized and that the isotropic model was an appropriate approach for all sites and years (bootstrap: $p >$ 0.22). According to the isotropic model, the estimated uphill and downhill mean dispersal distances at site B and E were 97 m (2015) and 86 m (2016) and 367 m (2015) and 380 m (2016), respectively (Table 3.3 and see left of Fig. 3.5).

Table 3.3 Inverse modelling results of isotropic and anisotropic dispersal (exponential function) of *Betula pendula* seeds for 2015 and 2016 (α - fecundity, λ - distance, β - coherency, ψ - rotation angle, γ - drift, rho - Spearman's correlation coefficient, p - *p*-value, φ - seed production of a single seed tree with a dbh of 20 cm).

Site	Year	Model	α	λ	β	ψ	γ	AIC	loglike	rho	*p*	φ
B	2015	Isotropic	2.02	96.73	-	-	-	509.92	-252.96	0.522	0.0001	301,169
		Anisotropic	2.15	430.60	0.065	-0.791	2.334	511.07	-250.54	0.602	0.0000	343,746
	2016	Isotropic	3.64	85.80	-	-	-	675.87	-335.93	0.636	0.0000	1,526,309
		Anisotropic	3.71	136.37	0.791	0.490	1.029	667.18	-328.59	0.804	0.0000	1,641,351
E	2015	Isotropic	2.21	367.08	-	-	-	329.37	-162.68	0.463	0.0023	366,028
		Anisotropic	1.30	234.89	0.005	0.888	185.673	337.16	-163.58	0.473	0.0018	146,696
	2016	Isotropic	3.57	379.77	-	-	-	413.72	-204.86	0.716	0.0000	1,427,139
		Anisotropic	2.14	124.03	0.839	0.600	2.143	420.50	-205.25	0.812	0.0000	339,410

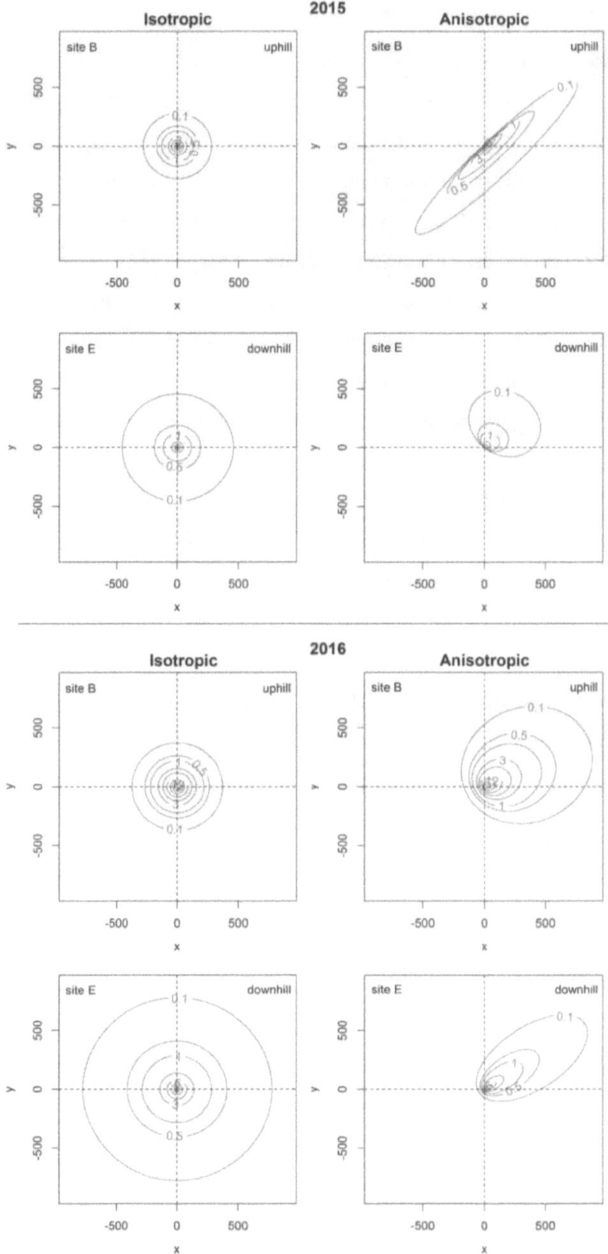

Fig. 3.5 Inverse model predictions of seed shadows [n m⁻²] simulated for a theoretical single *Betula pendula* seed tree with a dbh of 20 cm, given isotropic (left) and anisotropic (right) seed dispersal for study sites B and E in 2015 and 2016.

3.5 Discussion

3.5.1 Seed production

Rough knowledge about the number of seeds per seed tree and the quality of the seed crop in a particular year is important to forecast the probability of natural regeneration in disturbed areas. The amount of seeds produced by an individual tree is influenced by e.g. tree stem and crown dimension, vitality and age (Sarvas 1948, 1952; Moles et al. 2004) as well as by weather-dependent pollination and flowering success (Sarvas 1952).

The fecundity parameter α of the isotropic models was almost constant, which indicates a strong correlation between tree dbh and seed production, as previously assumed and confirmed for birch by Sato and Hiura (1998), Wagner et al. (2004) and Huth (2009). There is solid evidence for strict allometric relationships between growth parameters, such as dbh, crown radius or basal area of a tree, and the individual seed crop of a tree (Grisez 1975; Greene et al. 2004; Huth 2009; DaPonte Canova 2018). Therefore, the seed crop of birch trees of variable size can be easily estimated based on their dbh, if the seed production is not hampered by low viability or unfavourable weather conditions (see Grisez 1975). Birches with limited growing space, small crown projected areas and crown volumes have lower seed crops than large solitary individuals (Sarvas 1948). Nevertheless, Huth (2009) noted a restriction of these relationships. With the on setting senescence of a tree, its seed crop progressively decreases while its diameter and crown dimension generally continue to increase. The strength of the relationship therefore decreases with increasing tree age.

The predicted seed production of a single birch seed tree in 2016 was four times higher than in 2015. According to the findings of Sarvas (1948), 2016 can be described as good seed year. Birch is known for a large interannual variability in seed production (Sarvas 1948; Houle and Payette 1990; Kullman 1993; Huth 2009), which is mainly a response to the climatic conditions of the previous year (Kullman 1993; Holmström et al. 2017). On average, good seed years (i.e. so-called mast years) occur every three years (Sarvas 1948), during which the percentage of seed germination is significantly higher than in intervening (non-mast) years (Sarvas 1952; Bjorkbom 1971; Houle and Payette 1990). For a single silver birch, Denisow (2007, cited in Huth 2009) reported a seed production of 40,000-50,000 seeds in intervening years and 3.7-4.9 million seeds in a mast year. Compared to our study, significantly higher seed crops of individual trees with up to 7.3-10.0 million seeds (dbh of 24-80 cm) were reported by e.g., Arnborg (1948, cited in Perala and Alm 1990), Popadyuk et al. (1995), Wagner et al. (2004) and Huth (2009). In the present study, the slightly lower seed production of 0.6-6.5 million seeds per single seed tree (dbh of 13-42 cm) in a mast year might have resulted from smaller birch tree crowns due to strong spruce competition and a lack of release thinning in the past.

3.5.2 Directionality

Due to non-random anemochorous seed dispersal, previous studies often showed directionality for birch seed dispersal (Wagner et al. 2004; Wright et al. 2008; Huth 2009). However, this was not clearly confirmed by results of the present study. An explanation for the surprising isotropy of seed dispersal in this study might be the relatively long seed collection periods (four month). The variability of wind directions and wind speeds occurring during a period of four months – boosted and modified by turbulence and varying wind speeds in the open areas due to vegetation cover and structure (see Moon et al. 2013) – may explain the lack of observed anisotropy. While the half-hourly meteorological data (Table 3.1) showed no extreme wind events during the two study periods, with a maximum wind speed of 9.9 m s^{-1}, the monthly mean wind direction featured high standard deviations of ±25 to ±122°. On some days during the study periods, the variability of the wind direction was as high as ±206°. The observed isotropic distributions are thus plausible, because no prevailing wind direction was identified for both study periods and birch seeds were dispersed in a variety of directions. The assumption that the anisotropic models reveal a better fit to the empirical four-month data was therefore rejected. Houle and Payette (1990) found anisotropic spatial patterns of the seed shadows of *B. alleghaniensis* after subdividing the seed rain period into shorter study periods. Had we chosen shorter periods for emptying the seed traps in this study (e.g. 14 day periods), we might have also been able to detect anisotropy by inverse modelling (Wagner et al. 2004).

However, short-term analyses of 14 day periods would not be useful for deriving silvicultural recommendations, because in this context the entire period of seed rain, i.e. 3-4 months in summer and autumn, has to be considered. For silvicultural practice it is important to know that equally distributed seed rain can be expected around seed trees if strong wind regimes with variable wind directions prevail at a specific site.

3.5.3 Spatial patterns and seed dispersal distances

As expected, the highest seed densities where observed close to the seed trees. At study site B, where birch seed trees were positioned at the edge of the open area, higher densities were observed up to distances of 40-50 m. Similar results were reported by Sarvas (1948), Fries (1984), Skoglund and Verwijst (1989) and Cameron (1996). In our study, the trapped seed numbers at site B decreased rapidly at distances exceeding 50 m (see Fig. 3.4). The seed distribution thus showed a negative exponential seed dispersal kernel, as Bjorkbom (1971), Hughes and Fahey (1988), Greene and Johnson (1996) and Karlsson (2001) previously reported for *B. alleghaniensis*, *B. pendula*, *B. pubescence* and *B. papyrifera*. In theory, one may expect a log-normal distribution of the distances that the tiny and lightweight wind-dispersed seeds travel (Stoyan and Wagner 2001; Huth 2009), but the majority of records in this study showed no peaks at certain distances from seed trees (as in Greene and Johnson 1996; Stoyan and Wagner 2001; Huth 2009). A feasible explanation for this observation was given by Marquise (1969), who reported that primarily the heaviest and viable birch seeds were deposited close to the seed trees.

In the present study, seed dispersal distances within the sites were similar for both years regardless of extent of seed production. Large differences of dispersal distances only occurred between the study sites. The modelled mean isotropic dispersal distances (MDD_{iso}) of birch seeds distributed from the forest edge into the open area at site B were 86 and 97 m, as detected by Wagner et al. (2004) and Huth (2009) for birch in level closed forest stands. In contrast, the MDD_{iso} of 367 and 380 m modelled for seed dispersal from within the adjacen spruce forest stands into the larger open area E indicated dispersal distances that were four times larger. Hughes and Fahey (1988), Daniels (2001) and Karlsson (2001) recorded lower dispersal distances of 30-125 m for *B. alleghaniensis*, *B. pendula* and *B. pubescens* in open areas than observed in the present study.

Our very contradictory results of MDD are only comparable with studies from McEuen and Curran (2004), who found seed dispersal distances of *B. papyrifera* of 700 m between landscape fragments. The long distances were explained by an enormous seed tree presence (see also Zhao et al. 2016) and their extremely high seed production numbers. Similarly, we also had a huge seed source presence in the form of a mixed willow-rowan-birch stand at a distance of 74 m from the open area at site E. This potential explanation is further be confirmed by Greene and Johnson (1996), Zhao et al. (2016) and Holmström et al. (2017) who mentioned that seeds are dispersed over larger distances if the size of the canopy openings increases, if seed source densities are located close to forest edges or if seed tree density is high. Therefore, to increase the probability of getting birch seeds onto storm-felled sites, an increasing number of seed sources would be needed with longer distances to the respective sites. Nevertheless, the few birch seed trees around the open area at site B were standing at the forest edge, making it actually more likely that seeds would disperse over larger distances than in closed forests (Holmström et al. 2017).

Thus, another effect should be considered: the relief, although this effect has not often been mentioned in previous studies. In this study, it seems that the dispersal distances of birch seeds decreased uphill (site B) or increased downhill (site E) depending on the inclination and seed tree position (valley, slope or plateau) relative to the storm-felled site. Hill and Stevens (1981), who studied soil seed banks, found a 30 m shorter distance for uphill deposition of birch seeds than for downhill deposition. Based on the estimated MDD_{iso}, we may therefore assume a strong effect of inclination on the dispersal distance of birch seeds. However, the two study sites were not selected rigorously enough to test for inclination effects. Some of the other aforementioned factors, e.g. differently sized open areas, seed tree densities and distances between seed sources and forest edges, may also have influenced the dispersal distances.

The distance of long dispersal is particularly determined by secondary dispersal (Matlack 1989; Greene and Johnsons 1997), but this factor can be neglected in the present study due to similar ground vegetation cover and thus likely similar secondary dispersal in both open areas. Based on our data, we have no chance to check for secondary seed dispersal in any way.

3.6 Seed dispersal scenarios for silvicultural management decisions

3.6.1 Seed dispersal scenarios

The abundance and the spatial pattern of potential seed trees near a particular disturbed site are important determinants upon which forest managers could base silvicultural decisions about a risk-adapted reforestation concept and a preventative risk-adapted seed tree management.

Based on these results, scenarios of regular and aggregated seed tree distributions around the studied open areas were created to show that the distribution, the distance between seed trees and the disturbed sites, and the inclination have varying effects on the deposited seed densities. For analysing the scenarios, we could only consider the open areas, because the open area at site B represents a hilltop with only uphill seed dispersal (not the forested areas), and the sloped open area at site E forms a relief funnel which only represents downhill seed dispersal.

For an even, systematic seed tree distribution, the comparison of seed shadows between both sites showed higher deposited seed densities in the vicinity of trees at site B (Fig. 3.6a and c), because the same number of produced seeds are distributed uphill over smaller distances than downhill (see Fig. 3.5). However, in an undisturbed forest with a systematic seed tree distribution, there would be no parts without seed rain as the seed shadows of single seed trees would overlap at both sites. If disturbances interrupted the systematic seed tree grids – like observed at the studied storm-felled areas – and no seed trees were left in the disturbed area, approximately 20-25 n m⁻² birch seeds would still reach the hilltop (site B) and the valley (site E), independent of the inclination. It seems like there are no differences between sites, but the different sizes of the open areas must be considered when interpreting the scenarios. If the open area at site B were to exceed the size of the open area at site E, the seed densities deposited in the open area would be considerably lower, which illustrates the aggregated distribution of the same numbers of seed trees (Fig. 3.6b and d). At distances of 5-250 m between the seed trees and the forest edges, the seeds dispersed uphill with the shorter MDD are not able to reach the entire open area at site B, only the southern part. In contrast, the seed shadow of the seeds dispersed downhill from the aggregated seed trees at site E does not really differ from the spatial pattern of systematically distributed seed trees. At least 20 n m⁻² seeds are reaching almost all parts of the storm-felled area at site E with the exception of one small spot in the north.

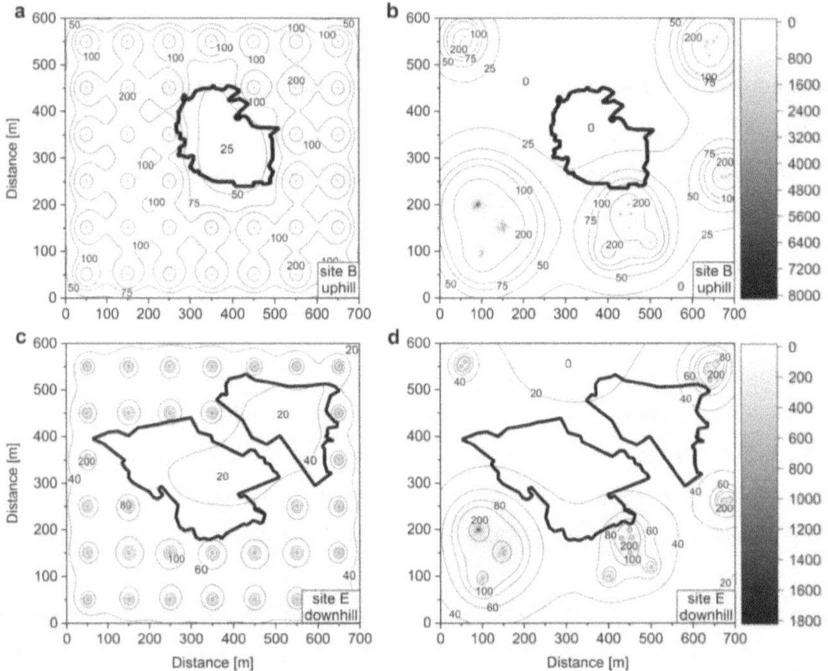

Fig. 3.6 Isotropic seed shadow scenarios [n m⁻²] of simulated systematic (left) and aggregated (right) seed tree distributions of *Betula pendula* around the studied open areas at sites B (top) and E (bottom), illustrated as bold lines. Note the different seed density scales for colour illustration in the figure panels.

The seed dispersal scenarios showed that in case of an even, systematic seed tree distribution, inclination differences are not clearly affecting the seed shadow, and known limits of birch dispersal distances (MDD) can generally be compensated, especially for uphill seed dispersal. However, if seed trees are aggregated and unevenly distributed, the effect of the inclination is clearly visible in the resulting seed shadow. In the latter case, the inclination has an important influence on regeneration success of storm-felled sites in addition to the distance between the seed sources and the open area as well as the number of seed trees or tree groups.

3.6.2 Conclusions for silvicultural management decisions

The findings of our case study indicated a strong influence of (i) site inclination and (ii) seed tree position (valley, slope or plateau) and distance to a storm-felled site on the seed dispersal of *B. pendula*. The birch seed shadow is also influenced by (iii) the number of seed sources. In the context of the natural regeneration of disturbed sites by *B. pendula*, it must be noted that seed dispersal is restricted compared e.g., to anemochorously dispersed *Salix* ssp. (Gage and Cooper 2005; Tiebel et al. 2019).

Willis et al. (2016) hence highlighted the importance of local seed source availability for successful birch regeneration. Sarvas (1948) mentioned that at least 100 to 200 n m^{-2} of viable seeds are necessary for successful regeneration and recommends a density of 4 to 8 n ha^{-1} of *B. pendula* seed trees, while Safford and Jacobs (1983) advocate a seed tree number of 7 to 12 n ha^{-1} for *B. papyrifera*. Although the relation between the seed numbers deposited on the ground and the actual number of established seedlings was not subject of the present study, it can be concluded from our results and model fits in combination with previously published studies that certain seed densities are not sufficient for successful regeneration. Regeneration success depends on many unpredictable conditions, which were not investigated in this study, and the mortality rate of birch seedlings can reach up to 99 % if germination conditions are unfavourable (Kinnaird 1974). For successful regeneration, birch seeds need microsites like bare ground with optimal moisture and light conditions (see Marquis 1966; Kinnaird 1974; Skoglund and Verwijst 1989; Karlsson 2001; Jonášová and Matějková 2007; Huth 2009; Willis et al. 2016).

However, based on Sarva's (1948) minimum recommendation of 100 seeds m^{-2}, we can give recommendations on the required seed tree numbers in forests. Derived from the uphill and downhill mean dispersal distances and the deposited seed numbers at study sites B and E (cf. Fig. 3.4 and 3.6), 4 and 16 n ha^{-1} seed trees are needed under unfavourable conditions, e.g. non-mast years, respectively. This corresponds to seed tree grid intervals of 60 and 30 m, respectively. For small groups of trees, the distance can be slightly larger, as illustrated in Fig. 3.6. Higher source numbers are needed for ensuring the natural regeneration of disturbed sites if the present seed trees produce insufficient seed crops as a result of lacking tending measures. Thus, the aforementioned seed tree numbers should be considered as minimum numbers.

For the practical management of regenerating disturbed sites, it can be concluded that seed dispersal is very sensitive to the distribution, the number and loss of seed trees in nearby forests as well as the size and inclination of disturbed areas. With respect to uphill dispersal, we assume that areas with insufficient numbers of deposited seeds will probably always occur in disturbed sites, unless the area is completely surrounded by seed trees. In case of disturbed sites of more than 4 ha, it is thus impossible to ensure a full cover of natural birch regeneration due to the limited seed dispersal distances. The natural regeneration of such large areas therefore should be supported by additional reforestation measures. In addition, seed trees surviving after a disturbance event must not be removed from the disturbed sites and seed trees in the vicinity of the sites should be promoted and vitalized wherever possible. Advance regeneration, which established underneath the canopy prior to the disturbance event, can also provide valuable benefits for regenerating disturbed sites. For the regeneration of small disturbed areas, which usually occur more frequently (Brang et al. 2015), a risk-adapted forest management should include the 'spatial optimization' of birch seed trees within conifer forests, due to the limited dispersal distance of birch seeds in general. A few groups of aggregated seed trees within a forest stand or some seed trees along the forest edge and forest roads in otherwise pure conifer forests are not sufficient for regenerating disturbed areas, but can be a good initial for the integration of birch trees in conifer forests. Along forest roads, paths or trails, birch trees have the possibility of unrestricted crown growth on one side and thus more proliferous seed production. However, a network of more or less regularly distributed individual birches is needed within conifer forests, preferably even of small groups of seed trees, because silvicultural measures are easier to implement (see Cameron 1996; Hynynen et al. 2010) and the chances of successfully regenerating more distant sites is higher. For birches, Cameron (1996) mentioned required thinning intervals of 5-7 years for ensuring a good crown growth. Conservation, vitalization and propagation are important factors for annual birch seed crop quality and quantity, which makes more sense for tree groups than for single trees due to costs and the regulation of interspecific competition.

3.7 References

Atkinson MD (1992) *Betula pendula* Roth (*B. verrucosa* Ehrh.) and *B. pubescens* Ehrh. J. Ecol. 80:837-870. 10.2307/2260870

Augspurger CK, Franson SE (1987) Wind dispersal of artificial fruits varying in mass, area, and morphology. Ecology 68:27-42. 10.2307/1938802

Bakker JP, Poschlod P, Strykstra RJ, Bekker RM, Thompson K (1996) Seed banks and seed dispersal: important topics in restoration ecology. Acta Bot. Neerl. 45:461-490. 10.1111/j.1438-8677.1996.tb00806.x

Beck P, Caudullo G, de Rigo D, Tinner W (2016) *Betula pendula, Betula pubescens* and other birches in Europe: distribution, habitat, usage and threats. In: San-Miguel-Ayanz J, de Rigo D, Caudullo G, Houston Durrant T, Mauri A (eds) European Atlas of Forest tree species. Luxembourg, pp 70-73.

Bjorkbom JC (1971) Production and germination of paper birch seed and its dispersal into a forest opening. Northeastern Forest Experiment Station, Forest Service Research Paper NE-209, U.S. Department of Agriculture, Upper Darby, USA

Brang P, Hilfiker S, Wasem U, Schwyzer A, Wohlgemuth T (2015) Langzeitforschung auf Sturmflächen zeigt Potenzial und Grenzen der Naturverjüngung. Schweiz. Z. Forstwes. 166:147-158.

Brouwer W, Stählin A (1975) Handbuch der Samenkunde für Landwirtschaft, Gartenbau und Forstwirtschaft. DLG-Verlags-GmbH, Frankfurt (Main).

Bürger M (2003) Bodennahe Windverhältnisse und windrelevante Reliefstrukturen. In: Leibnitz-Institut für Länderkunde (ed) Nationalatlas Bundesrepublik Deutschland. Spektrum Akademischer Verlag, pp 52-55

Burse K-D, Schramm H-J, Geiling S, Meinhardt H, Schölch M (1997) Die forstlichen Wuchsbezirke Thüringens - Kurzbeschreibung, Mitteilungen der Landesanstalt für Wald und Forstwirtschaft, Gotha.

Bushart M, Suck R (2008) Potenzielle natürliche Vegetation Thüringens, Schriftenr. der Thüringer Landesanstalt für Umwelt und Geologie, Jena.

Cameron AD (1996) Managing birch woodlands for the production of quality timber. Forestry 69:357-371. https://doi.org/10.1093/forestry/69.4.357

Clark JS, Silman M, Kern R, Macklin E, HilleRisLambers J (1999) Seed dispersal near and far: patterns across temperate and tropical forests. Ecology 80:1475-1494. 10.1890/0012-9658(1999)080[1475:SDNAFP]2.0.CO;2

DaPonte Canova G (2018) Regeneration ecology of anemochorous tree species *Qualea grandiflora* Mart. and *Aspidosperma tomentosum* Mart. of the cerrado Aguara Ñu located in the Mbaracayú Nature Forest Reserve (MNFR), Paraguay. book, Technical University of Dresden

Daniels J (2001) Ausbreitung der Moorbirke (*Betula pubescens* Ehrh. agg.) in gestörten Hochmooren der Diepholzer Moorniederung. Osnabrücker Naturwissenschaftliche Mitteilungen 27:39-49

Dorp D van, Hoek WPM van den, Daleboudt C (1996) Seed dispersal capacity of six perennial grassland species measured in a wind tunnel at varying wind speed and height. Can J Bot 74:1956-1963. https://doi.org/10.1139/b96-234

Faraway JJ (2006) Extending the linear model with R - Generalized linear mixed effects and nonparametric regression models. Chapman & Hall/CRC, Boca Raton, London, New York

Fenner M (1985) Seed ecology. Chapman and Hall Ltd, London and New York

Ferris R, Humphrey JW (1999) A review of potential biodiversity indicators for application in British forests. Forestry 72:313-328. https://doi.org/10.1093/forestry/72.4.313

Fink AH, Brücher T, Ermert V, Krüger A, Pinto JG (2009) The European storm Kyrill in January 2007: synoptic evolution, meteorological impacts and some considerations with respect to climate change. Nat. Hazards Earth Syst. Sci. 9:405-423. 10.5194/nhess-9-405-2009

Fries G (1984) Den frösådda björkens invandring på hygget (Immigration of birch into clearfelled areas). Sver. Skogsvardsfdrb. Tidskr. 82:35-39

Frischbier N, Profft I, Hagemann U (2014) Potential impacts of climate change on forest habitats in the Biosphere Reserve Vessertal-Thuringian Forest in Germany. In: Rannow

S, Neubert M (eds.) Managing protected areas in Central and Eastern Europe under climate change, Advances in Global Change Research. Springer Science and Business Media, Dordrecht, pp 243-257. https://doi.org/10.1007/978-94-007-7960-0_16

Gage EA, Cooper DJ (2005) Patterns of willow seed dispersal, seed entrapment, and seedling establishment in a heavily browsed montane riparian ecosystem. Can. J. Bot. 83:678-687. https://doi.org/10.1139/B05-042

Gauer J, Aldinger E (2005) Waldökologische Naturräume Deutschlands: Forstliche Wuchsgebiete und Wuchsbezirke - mit Karte 1:1.000.000, Mitteilungen des Vereins für Forstliche Standortskunde und Forstpflanzenzüchtung, Stuttgart

Graber RE, Leak WB (1992) Seed fall in an old-growth northern hardwood forest. Northeastern Forest Experiment Station, Forest Service Research Paper NE-663, Pennsylvania, USA

Greene DF, Johnson EA (1996) Wind dispersal of seeds from a forest into a clearing. Ecology 77:595-609. 10.2307/2265633

Greene DF, Johnson EA (1997) Secondary dispersal of tree seeds on snow. J. Ecol. 85(3):329-340. 10.2307/2960505

Greene DF, Canham CD, Coates KD, Lepage PT (2004) An evaluation of alternative dispersal functions for trees. J. Ecol. 92:758-766. https://doi.org/10.1111/j.0022-0477.2004.00921.x

Gregow H, Laaksonen A, Alper ME (2017) Increasing large scale windstorm damage in Western, Central and Northern European forests, 1951–2010. Sci. Rep. 7:1-7. 10.1038/srep46397

Grisez T (1975) Flowering and seed production in seven hardwood species. Northeastern Forest Experiment Station, Forest Service Research Paper NE-315, Upper Darby, USA

Hill MO, Stevens PA (1981) The density of viable seed in soils of forest plantations in upland Britain. J. Ecol. 69:693-709. 10.2307/2259692

Holmström E, Karlsson M, Nilsson U (2017) Modelling birch seed supply and seedling establishment during forest regeneration. Ecol. Model. 352:31-39. http://dx.doi.org/10.1016/j.ecolmodel.2017.02.027

Houle G, Payette S (1990) Seed dynamics of *Betula alleghaniensis* in a deciduous forest of north- eastern North America. J. Ecol. 78:677-690. 10.2307/2260892

Houle G (1998) Seed dispersal and seedling recruitment of *Betula alleghaniensis*: Spatial inconsistency in time. Ecology 79:807-818. https://doi.org/10.1890/0012-9658(1998)079[0807:SDASRO]2.0.CO;2

Hughes JW, Fahey TJ (1988) Seed dispersal and colonization in a disturbed northern hardwood forest. Bull. Torrey Bot. Club 115: 89-99

Humphrey JW, Holl K, Broome AC (1998) Birch in spruce plantations - Management for Biodiversity. Forestry Commission Technical Paper 26

Huth F (2009) Untersuchungen zur Verjüngungsökologie der Sand-Birke (*Betula pendula* Roth). book, Technical University of Dresden

Hynynen J, Niemistö P, Viherä-Aarnio A, Brunner A, Hein S, Velling P (2010) Silviculture of birch (*Betula pendula* Roth and *Betula pubescens* Ehrh.) in northern Europe. Forestry 83:103-119.10.1093/forestry/cpp035

Jonášová M, Matějková I (2007) Natural regeneration and vegetation changes in wet spruce forests after natural and artificial disturbances. Can. J. For. Res. 37:1907-1914. 10.1139/X07-062

Kadereit JW, Leins P (1988) A wind tunnel experiment on seed dispersal in *Papaver* L. sects. *Argemonidium* SPACH and *Rhoeadium* SPACH (*Papaveraceae*). Flora 181:189-203. 10.1016/S0367-2530(17)30365-1

Karlsson M (2001) Natural regeneration of broadleaved tree species in southern Sweden: effects of silvicultural treatments and seed dispersal from surrounding stands. book, Swedish University of Agricultural science

Kinnaird JW (1974): Effect of site conditions on the regeneration of birch (*Betula pendula* Roth and *B. pubescens* (Ehrh.). J. Ecol. 62:467-472. 10.2307/2258992

Koski V, Rousi M (2005) A review of the promises and constraints of breeding silver birch (*Betula pendula* Roth) in Finland. Forestry 78:187-198. 10.1093/forestry/cpi017

Kullman L (1993) Tree limit dynamics of *Betula pubescens* ssp. *tortuosa* in relation to climate variability: evidence from central Sweden. J. Veg. Sci. 4:765-772. https://doi.org/10.2307/3235613

Leder B (1992) Weichlaubhölzer: Verjüngungsökologie, Jugendwachstum und Bedeutung in Jungbeständen der Hauptbaumarten Buchen und Eiche, Schriftenreihe der Landesanstalt für Forstwissenschaft Nordrhein-Westfalen

Luostarinen K, Verkasalo E (2000) Birch as sawn timber and in mechanical further processing in Finland. A literature study. Silva Fennica Monographs 1

Marquis DA (1966) Germination and growth of paper birch and yellow birch in simulated strip cuttings. Northeastern Forest Experiment Station, Forest Service Research Paper Ne-54, Upper Darby, USA

Marquise DA (1969) Silvical requirements for natural birch regeneration. Northeastern Forest Experiment Station, Forest Service, Upper Darby, USA

Matlack GR (1989) Dispersal of seed across snow in *Betula lenta*, a gap-colonizing tree species. J. Ecol. 77:853-869. 10.2307/2260990

McEuen AB, Curran LM (2004) Seed dispersal and recruitment limitation across spatial scales in temperate forest fragments. Ecology 85:507-518. https://doi.org/10.1890/03-4006

Moles AL, Falster DS, Leishman MR, Westoby M (2004) Small-seeded species produce more seeds per square metre of canopy per year, but not per individual per lifetime. J. Ecol. 92:384-396. https://doi.org/10.1111/j.0022-0477.2004.00880.x

Mölter T, Schindler D, Albrecht AT, Kohnle U (2016) Review on the projections of future storminess over the North Atlantic European region. Atmosphere 7:1-40. 10.3390/atmos7040060

Okubo A, Levin SA (1989) A theoretical framework for data analysis of wind dispersal of seeds and pollen. Ecology 70:329-338. 10.2307/1937537

Patterson GS (1993) The value of birch in upland forests for wildlife conservation. Forestry Commission Bulletin 109

Perala DA, Alm AA (1990) Reproductive ecology of birch: A review. For. Ecol. Manag. 32:1-38. https://doi.org/10.1016/0378-1127(90)90104-J

Popadyuk RV, Smirnova OV, Evstigneev OI, Yanitskaya TO, Chumatchenko SI, Zaugolnova LB, Korotkov VN, Chistyakova AA, Khanina LG, Komarov AS (1995) Current state of broad-leaved forests in Russia, Belorussia, Ukraine: historical development, biodiversity, structure and dynamic. Preprint, PRC RAS, Pushchino

Priha O (1999) Microbial activities in soils under Scots pine, Norway spruce and silver birch. book, University of Helsinki

Putten B van, Visser MD, Muller-Landau HC, Jansen PA (2012) Distorted-distance models for directional dispersal: a general framework with application to a wind-dispersed tree. Methods Ecol. Evol. 3:642-652. https://doi.org/10.1111/j.2041-210X.2012.00208.x

R Core Team (2014) R: A Language and Environment for Statistical Computing. R Foundation for Statistical Computing, Vienna. https://www.R-project.org

Ribbens E, Silander JA Jr, Pacala SW (1994) Seedling recruitment in forest: calibrating models to predict patterns of tree seedling dispersion. Ecology 75:1794-1806. 10.2307/1939638

Röhrig E, Gussone HA (1990) Waldbau auf ökologischer Grundlage. Band 2: Baumartenwahl, Bestandesbegründung und Bestandespflege. Verlag Paul Parey, Hamburg - Berlin

Roloff A, Pietzarka U (2010) *Betula pendula* ROTH. In: Roloff A, Weisberger H, Lang U, Stimm B (eds) Bäume Mitteleuropas. WILEY-VCH Verlag GmbH & Co. KGaA, Weinheim, pp 61-75

Safford LO, Jacobs RD (1983) Paper birch. In: Burns RM (eds) Silvicultural systems for the major forest types of the United States, U.S. Department of Agriculture Washington, D. C., Agriculture Handbook No. 445, pp 145-147

Sarvas R (1948) A research on the regeneration of birch in South Finland. Commun. Inst. For. Fenn. 35:82-91

Sarvas R (1952) On the flowering of birch and the quality of seed crop. Commun. Inst. For. Fenn. 40:1-35

Sato H, Hiura T (1998) Estimation of overlapping seed shadows in a northern mixed forest. For. Ecol. Manag. 104:69-76. https://doi.org/10.1016/S0378-1127(97)00247-8

Skarpaas O, Stabbetorp OE, Rønning I, Svennungsen TO (2004) How far can a hawk's beard fly? Measuring and modelling the dispersal of *Crepis praemorsa*. J. Ecol. 92:747-757. 10.1111/j.0022-0477.2004.00915.x

Skarpaas O, Auhl R, Shea K (2006) Environmental variability and the initiation of dispersal: turbulence strongly increases seed release. Proc. Biol. Sci. 273:751-756. https://doi.org/1098/rspb.2005.3366

Skoglund J, Verwijst T (1989) Age structure of woody species populations in relation to seed rain, germination and establishment along the river Dalälven, Sweden. Vegetation 82:25.34. https://doi.org/10.1007/BF00217979

Soubeyrand S, Enjalbert J, Sanchez A, Sache I (2007) Anisotropy, in density and in distance, of the dispersal of yellow rust of wheat: experiments in large field plots and estimation. Phytopathology 97:1315-1324. https://doi.org/10.1094/PHYTO-97-10-1315

Stoyan D, Wagner S (2001) Estimating the fruit dispersion of anemochorous forest trees. Ecol. Model. 145:35-47. https://doi.org/10.1016/S0304-3800(01)00385-4

Tekle FB, Gudicha DW, Vermunt JK (2016) Power analysis for the bootstrap likelihood ratio test for the number of classes in latent class models. Adv Data Anal Classif 10:209-224. 10.1007/s11634-016-0251-0

Tiebel K, Leinemann L, Hosius B, Schlicht R, Frischbier N, Wagner S (2019) Seed dispersal capacity of *Salix caprea* L. assessed by seed trapping and parentage analysis. Eur J For Res. 138:495-511. https://doi.org/10.1007/s10342-019-01186-2

Waesch G (2003) Montane Graslandvegetationen des Thüringer Waldes: Aktueller Zustand, historische Analyse und Entwicklungsmöglichkeiten. Cuvillier Verlag, Göttingen

Wagner S, Wälder K, Ribbens E, Zeibig A (2004) Directionality in fruit dispersal models for anemochorous forest trees. Ecol. Model. 179:487-498. https://doi.org/10.1016/j.ecolmodel.2004.02.020

Wälder K, Näther W, Wagner S (2009) Improving inverse model fitting in trees-Anisotropy, multiplicative effects, and Bayes estimation. Ecol. Model. 220:1044-1053. https://doi.org/10.1016/j.ecolmodel.2009.01.034

Willis JL, Walters MB, Farinosi E (2016) Local seed source availability limits young seedling populations for some species more than other factors in northern hardwood forests. For. Sci. 62:440-448. http://dx.doi.org/10.5849/forsci.15-143

Wright SJ, Trakhtenbrot A, Bohrer G, Detto M, Katul GG, Horvitz N, Muller-Landau HC, Jones FA, Nathan R (2008) Understanding strategies for seed dispersal by wind under contrasting atmospheric conditions. PNAS 105:19084-19089. 10.1073pnas.0802697105

Zar JH (2010) Biostatistical analysis. Prentice Hall, Upper Saddle River, New Jersey

Zerbe S (2001) On the ecology of *Sorbus aucuparia* (*Rosaceae*) with special regard to germination, establishment and growth. Pol Bot J 46:229-239. 10.1234/12345678

Zerbe S (2009) Renaturierung von Waldökosystemen. In: Zerbe S, Wiegleb G (eds) Renaturierung von Ökosystemen in Mitteleuropa. Spektrum Akademischer Verlag, Heidelberg, pp 153-182

Zhao F, Qi L, Fang L, Yang J (2016) Influencing factors of seed long-distance dispersal on a fragmented forest landscape on Changbai Mountains, China. Chin. Geogra. Sci. 26:68-77. 10.1007/s11769-015-0747-0

Chapter 4

The impact of structural elements on storm-felled sites on endozoochorous seed dispersal by birds – a case study

Katharina Tiebel[1], Franka Huth[1], Alexandra Wehnert[1], Sven Wagner[1] (under review)

iForest

[1]TU Dresden, Institute of Silviculture and Forest Protection, Chair of Silviculture, Pienner Str. 8, 01737 Tharandt, Germany

4.2 Introduction

Monospecific, even-aged and large scale Norway spruce forests at moderately high elevations in central Europe are prone to severe disturbances brought about, for example, by snow, storm and bark beetles (Löf et al. 2010, Profft 2013). As a consequence, artificial reforestation is a seemingly unavoidable, and costly, implication of this kind of forest management. Often, however, forest companies lack the financial and human resources required to ensure the timely clearance from sites of damaged timber and the reforestation of the disturbed areas. A cheaper alternative is natural succession by pioneer tree species.

Pioneer forests in central Europe consist of tree species like *Betula* ssp., *Salix* ssp., *Populus* ssp. and *Sorbus aucuparia* L., which can colonise damaged areas rapidly after disturbance due to their regular fructification and capacity for large seed dispersal distances. Pioneer trees are able to mitigate the negative consequences associated with these open areas quickly, provided there are sources of pioneer seed trees in the vicinity of disturbed areas. Over time, pioneer tree species may also compensate the negative ecological effects caused by homogenous Norway spruce and Scots pine forests and act as lasting stabilising structural elements. These species generally also enhance forest biodiversity. Therefore, pioneer tree species have a high ecological relevance within the aforementioned ecosystems (see Kay 1985, Perala & Alm 1990, Leder 1992, Schmidt 1998, 1999, Hacker 1999, Raspé et al. 2000, Kuzovkina & Quigley 2005, Argus 2006, Hynynen et al. 2010, Zerbe 2009).

Pioneer tree species differ in relation to their seed distribution mechanisms. Seeds can be dispersed by wind ('anemochorous'), water ('hydrochorous') or animals ('zoochorous') (see McVean 1956, Perala & Alm 1990, Worrell 1995, Raspé et al. 2000). The seed dispersal of anemochorous pioneer tree species has been investigated in numerous studies (see Leder 1992, Karlsson 2001, Wagner et al. 2004, Huth 2009). The mechanisms of zoochorous – and especially of endozoochorous – dispersal and its ecological and economic contribution to the regeneration of disturbed forest areas are still poorly understood (cf. McDonnell & Stiles 1983, McDonnell 1986, Hoppes 1987, Jordano & Schupp 2000, Stiebel 2003, Albrecht et al. 2012).

Endozoochorous seed dispersal, or dispersal after the passage of seed through the animal digestive tract, is carried out mainly by birds (e.g., blackbird, starling and thrush) and small mammals (e.g., dormouse, bank vole, brown vole and squirrel) (Erlbeck 1998, Schmidt 1998, Paulsen & Högstedt 2002). After consumption of the seed-bearing fruit, the undamaged seeds are deposited elsewhere by the animals (Bakker et al. 1996, Paulsen & Högstedt 2002). As small mammals often have a small dispersal range, the distribution capability is limited (Bakker et al. 1996, Kollmann 2000). Birds provide for the largest seed dispersal distance in endozoochory, for example, of rowan seed (*Sorbus aucuparia* L.).

An important aspect in the dispersal behaviour of birds is that the perches or habitats where defecation takes place are not chosen randomly (Obeso et al. 2011). Birds generally prefer to rest within protective forest edges or in more or less closed forest (Gregor & Seidling 1997, Jordano & Schupp 2000, Stiebel 2003, Żywiec & Ledwoń 2008, Żywiec 2014), although bird species differ in their habitat requirements (Stiebel 2003, Albrecht et al. 2012). Open areas without structural elements such as young trees, root plates, standing deadwood and stumps are largely avoided by birds, as these bare sites offer no perches and resting places or protection against predators (McDonnell & Stiles 1983, Stimm & Böswald 1994, Stiebel 2003, Żywiec & Ledwoń 2008). However, local fruit availability in open areas attracts many bird species (McDonnell & Stiles 1983, Albrecht et al. 2012). The distribution of the seed of endozoochorously dispersed tree species by frugivorous birds may, therefore, be hampered by structural limitations on disturbed forest sites.

To predict endozoochorous seed dispersal on open areas, knowledge of the post-foraging behaviour of bird species and of the characteristics of the open area is necessary. By using records of deposited droppings one can obtain knowledge of the behaviour of fruit-consuming bird species (McDonnell & Stiles 1983, García et al. 2007, Guitian & Munilla 2010). In 2015 we studied the spatial patterns of bird droppings on windthrown forest sites in Thuringia, Germany. We adopted a case study approach as this represents a good means of studying complex phenomena in ecology and so enhancing understanding of and underpinning existing general ecological theory (Baxter & Jack 2008). The aim of the case study presented herein was to answer the following three questions:

i)   Are there more bird droppings in dropping traps under structural elements on windthrown sites than in dropping traps on open areas without these elements?

ii)   Are certain structural elements preferred by frugivorous birds on open areas?

iii) Is the height of a structural element an important factor that can serve to determine the relevance of an element as a dropping site?

4.3 Materials and methods

4.3.1 Study area

The chosen study area is located at high elevations and along the ridges of the Thuringian Forest region, which is a mountain range in the German federal state Thuringia (50°40'N and 10°45'E). It is situated between 400-982 m above sea level (a.s.l.), with a prevailing south-westerly exposition. The region is characterised by many slopes and a near absence of plateaus (Burse et al. 1997, Waesch 2003, Gauer & Aldinger 2005). The mean annual precipitation ranges from 800 mm in the southwest to 1,200 mm along the ridges and falls to 700 mm in the northeast (Burse et al. 1997, Gauer & Aldinger 2005, Bushart & Suck 2008). The annual average temperature in the region varies between 4-6 °C (Burse et al. 1997, Bushart & Suck 2008). The area is influenced by an Atlantic, moderately cool and moist central mountain climate (Burse et al. 1997, Gauer & Aldinger 2005). The dominant soil types of the forest sites are cambisols with low levels of base saturation and low to medium nutrient contents (Gauer & Aldinger 2005). The landscape features a largely contiguous forest system of approximately 100,000 ha, with ~90 % forest cover, some small upland meadows in stream valleys and occasional small raised bogs. The study area is dominated by single-layered, even-aged Norway spruce forests (*Picea abies* (L.) Karst.), whereas the predominant potential natural vegetation types would be dominated by *Luzulo-Fagetum* and *Asperulo-Fagetum* beech forests (Frischbier et al. 2014).

We selected five study sites (A-E) located on slopes and mountain tops (plateaus) at higher elevations and near the ridges of the Thuringian Forest (715-900 m a.s.l.). Each site consisted of an open area surrounded by Norway spruce forest stands. All open areas originated with the storm 'Kyrill' in January 2007 (Fink et al. 2009). Before the storm, the sites were dominated by 68-100 year-old Norway spruce stands. After the storm, the damaged areas were completely cleared, and no rowan seed trees were present in the open areas. The size of the open areas on these study sites ranged from 4.0-12.7 ha (Table 4.1), and only limited tree regeneration had occurred. The surrounding forest stands were also dominated by 59-122 year-old Norway spruce forests, admixed with a small number of isolated birch (*Betula pendula* Roth), willow (*Salix caprea* L.) and rowan (*Sorbus aucuparia* L.) trees of a comparable age.

Table 4.1 Characteristics of the study sites and the experimental design.

Study sites	A	B	C	D	E
Elevation above sea level [m]	845–900	735–765	840–880	865–895	715–775
Topography	mountain peak with one slope	mountain peak with slopes	flat area	mountain peak with slopes	slopes
Size of open area [ha]	5.98	4.03	7.46	5.59	12.70
Experimental design of dropping trap positions	2 crossing line transects	4 crossing line transects	2 crossing line transects	4 crossing line transects	2 crossing line transects
Number of dropping traps in neighbouring spruce forests [n]	5	13	6	7	7
Number of dropping traps located near structural elements [n]	23	43	32	51	34
Number of dropping traps without connection to structural elements [n]	23	43	32	48	33

4.3.2 Experimental design

Dropping traps were used to catch bird droppings on the windthrown sites. The traps were placed in 2015, along either two or four crossing line transects with intervals of 20 m between the traps, covering the open areas on each study site. The number, orientation and length of the line transects were individually adapted to the expanse and shape of each study site. The line transects were extended over the entire open area of each study site and into the surrounding spruce forests (Fig. 4.1).

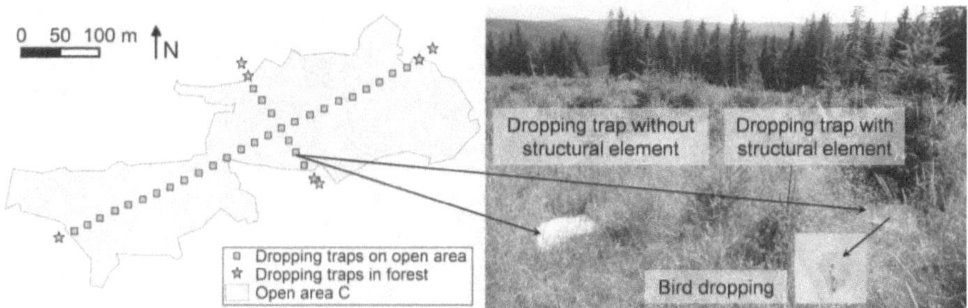

Fig. 4.1 Example of the experimental study design on study site C with two crossing line transects.

The dropping traps consisted of textile pieces covering a surface area of 0.25 m² and affixed to the soil surface with hooks. Rainwater could percolate through the textile while bird droppings remained on the textile surface. For each marked point along the line transects (see Fig. 4.1), one trap was placed on the ground in a location without a structural element within 2 m of the marked point (category: 'without structural element'), and a second trap was placed directly next to a structural element located within a maximum radius of 10 m from the marked point (category: 'with structural element') (see Fig. 4.1). The total of 179 traps without structural elements on open sites was lower than the 183 traps on open sites with structural elements. The reason for this was the high number of structural elements present on the study sites. The 38 additional dropping traps located in the neighbouring spruce forests (category: 'forest') were placed close to the base of the stems of old spruce trees.

The four main categories of structural element with the potential to serve as resting places for birds on open areas were as follows: (i) deadwood (dead *Fagus sylvatica* saplings, towering dead branches, upturned root plates, low and high stumps), (ii) established young trees of the species Norway spruce, silver birch, rowan and European beech, (iii) artificially introduced elements (game fencing and individual tree protectors), and (iv) old spruces on forest edges (Fig. 4.2). The heights of the selected structural elements ranged between 0.35 m and 25 m.

This variety of available structural elements allowed for study of the endozoochorous dispersal of seed by frugivorous birds on the five storm-felled study sites. The variation in the types, heights and numbers of the chosen structural elements is presented in Fig. 4.2 and Table 4.2. All dropping traps were mapped using a blumax Bluetooth GPS-4013 receiver. The traps were checked and cleaned every 3 or 4 weeks from July to November 2015, and only droppings of frugivorous birds were counted.

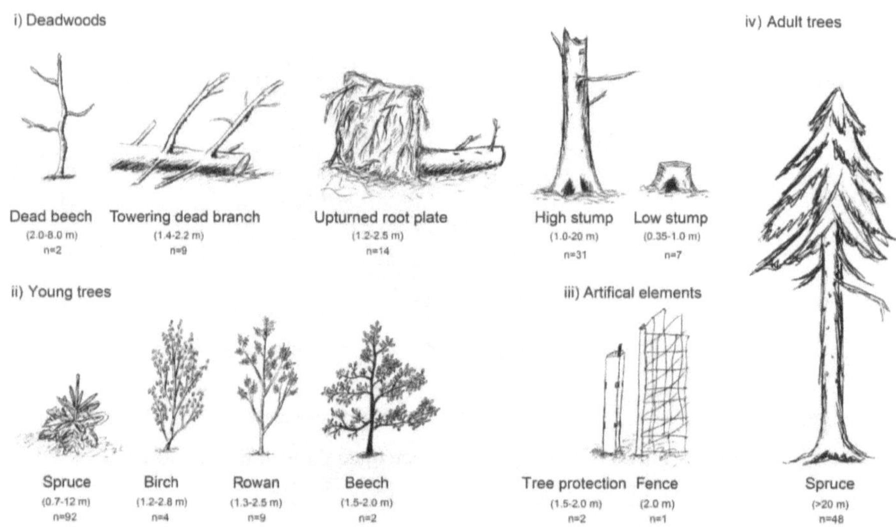

Fig. 4.2 The categories of structural element considered in the study differentiated by type, height (in brackets) and number (n).

4.3.3 Data analysis

The mean bird dropping density per m² was calculated for each dropping trap. Differences between the bird dropping densities between the study sites and the structural elements were analysed using the Kruskal-Wallis H-test as the data were not normally distributed. Where significant differences were ascertained ($p < 0.05$), the Mann-Whitney U-test was applied with a Bonferroni correction as an adjustment method to obtain additional information about the groups of differences. The Spearman and Pearson correlation coefficient (*rho*) was used to test association between area sizes or trap numbers with recorded bird dropping densities (Zar 2010). The statistical analyses were conducted using the R software version 3.3.2 (R Core Team 2014).

4.4 Results

The dropping traps situated within the categories 'without structural element', 'with structural element' and 'forest' represented 45 % (179 traps), 46 % (183 traps) and 9 % (38 traps) of all 400 traps installed, respectively.

The numbers of each the various types of structural element occurring within plots of the category 'with structural element' differed quite strongly for the 183 dropping trap locations situated on open areas (Table 4.2). The most frequently used structural element on open sites was young spruce trees, accounting for 50 % of all 183 samples. The frequencies of the dropping traps connected to young trees of other species as the corresponding structural elements were 5 % rowan, 2 % birch and 1 % beech. Dropping traps next to old spruce trees on forest edges accounted for 6 % of the 183 dropping trap locations situated in open areas under structural elements. The deadwood elements category accounted for 34 % of all 183 traps with structural elements in open areas, including 17 % high stumps, 8 % upturned root plates, 5 % towering dead branches, 4 % low stumps and 1 % dead beech.

Table 4.2 Characterisation of recorded structural element heights and nearby bird droppings [n m^{-2}] on all study sites A-E (N_d - number of recorded droppings, N_{st} - number of studied structural elements, max - maximum, min - minimum, sd - standard deviation).

Structural element category	N_{st}	Frequency	Element height [m]			N_d	Bird droppings [n m^{-2}]			
	Σ	[%]	min	mean	max	Σ	min	mean	max	± sd
With structural element	*183*	*100*								
Deadwood										
Dead beech	2	1.1	2.0	5.0	8.0	24	4	12.0	20	± 11.3
Towering dead branch	9	4.9	1.4	1.7	2.2	180	0	20.0	84	± 30.8
Root plate	14	7.7	1.2	1.7	2.5	64	0	4.6	48	± 12.6
High stump	31	16.9	1.1	3.5	20.0	120	0	3.9	56	± 10.6
Low stump	7	3.8	0.4	0.7	1.0	4	0	0.6	4	± 2.9
Young trees										
Spruce	92	50.3	0.7	2.8	12	96	0	1.1	16	± 2.9
Birch	4	2.2	1.2	2.0	2.8	0	0	0	0	-
Rowan	9	4.9	1.3	1.8	2.5	0	0	0	0	-
Beech	2	1.1	1.5	1.8	2.0	12	0	6.0	12	± 8.5
Artificial elements	3	1.6	1.8	1.9	2.0	0	0	0	0	-
Old spruce trees (forest edge)	10	5.5	> 20	> 20	> 20	17	0	1.7	12	± 3.6

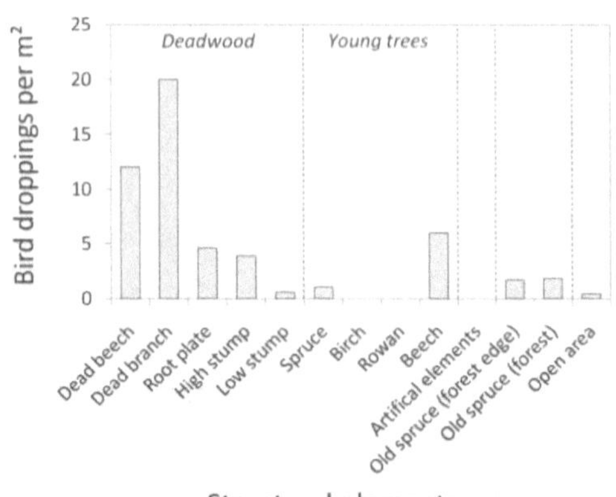

Fig. 4.3 Aggregate bird droppings [n m^{-2}] for each dropping trap position, including all study sites.

A total of 166 bird droppings were recorded on the 400 traps, corresponding to an average density of 0.42 droppings per m². However, whereas bird droppings were collected from 23 % of the 183 dropping traps situated under structural elements on open sites, only 8 % of the 179 traps 'without structural elements' produced droppings. High bird dropping densities were found under deadwood elements, but the frequency of droppings varied greatly between elements within the deadwood category (Fig. 4.3). The highest mean bird dropping density was collected under towering dead branches with 20 n m^{-2}, followed by upturned root plates (4.6 n m^{-2}), high stumps (3.9 n m^{-2}) and finally low stumps (0.6n m^{-2}). Dropping traps situated under artificial elements, and under young birch and rowan trees received no droppings. An average of 1.1 droppings per m² were found under young spruce and 1.7 n m^{-2} on traps near old spruce trees at forest edges. In the neighbouring spruce stands (category: 'forest') the mean dropping density was 1.8 n m^{-2} (Table 4.2). The more frequent use of certain structural elements (N_{st}) was not associated with a greater probability of the sampling of bird droppings (N_d) on these dropping traps (Pearson correlation: $r = 0.44$, $p = 0.173$).

A comparison of the mean dropping densities in the categories 'forest' (1.7 n m^{-2}) and 'with structural elements' (2.9 n m^{-2}) with the category 'without structural elements' (0.4 n m^{-2}) revealed significant differences (Kruskal-Wallis H-test: $p = 0.0001$ - Fig. 4.4).

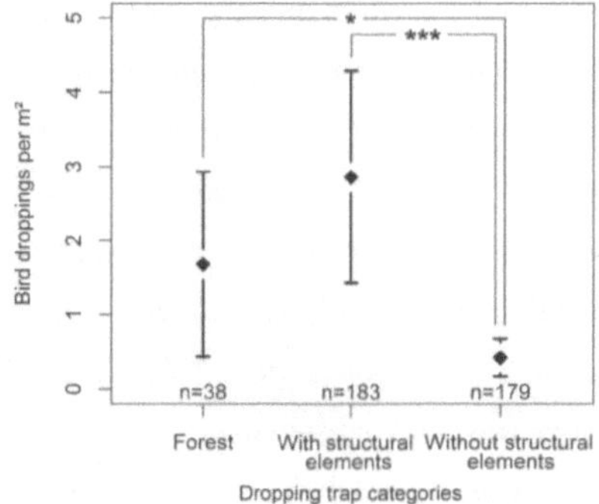

Fig. 4.4 Mean bird droppings [n m⁻²] recorded for the three categories of dropping trap position, including all study sites (Mann-Whitney U-test: $p < 0.001$).

With the exception of site E, the recorded mean densities of bird droppings on each open area were higher on traps under structural elements than on traps 'without structure elements', increasing from 0.5 n m⁻² on site A to 7.9 n m⁻² on site C (Fig. 4.5). Considerably smaller densities were found on dropping traps 'without structural elements', with the mean droppings ranging from 0.3 n m⁻² (site D) to 0.7 n m⁻² (site B). On the traps in the neighbouring forests an average of 0-2.8 droppings per m² were sampled. Although no faeces were detected on the traps on the largest study site (site E, area: 12.7 ha), no negative correlation was found between the site size and the recorded dropping densities per category (without structural element, with structural element, forest) independent of the occurrence of structural elements (Spearman correlation: $r < -0.17, p > 0.096$).

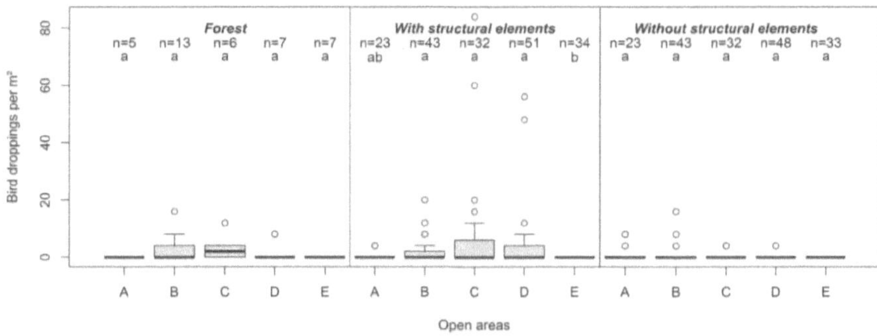

Fig. 4.5 Bird droppings [n m⁻²] for each of the study sites A-E and the three categories of dropping trap position: 'forest' (left), 'with structural elements' (middle) and 'without structural elements' (right). Letters indicate significant differences between the areas A-E (Mann-Whitney U-test: $p < 0.05$).

The preference of frugivorous birds for different heights of the structural elements considered was also assessed. Faeces of frugivorous birds were only recorded on dropping traps located near structural elements higher than 1 m in height. Further analysis was done for the largest and tallest categories of structural element: young trees and deadwood. As can be seen in Fig. 4.6, birds used all available deadwoods from 1-20 m in height, while droppings in the vicinity of young trees were only found under individuals between 1-4 m in height. No bird droppings were recorded on traps situated under young trees > 4 m. No correlation between element heights and dropping densities was detected for either of the above element categories, or the structural elements in general (Spearman correlation: $r < 0.1$ and $p > 0.1$ - Fig. 4.6).

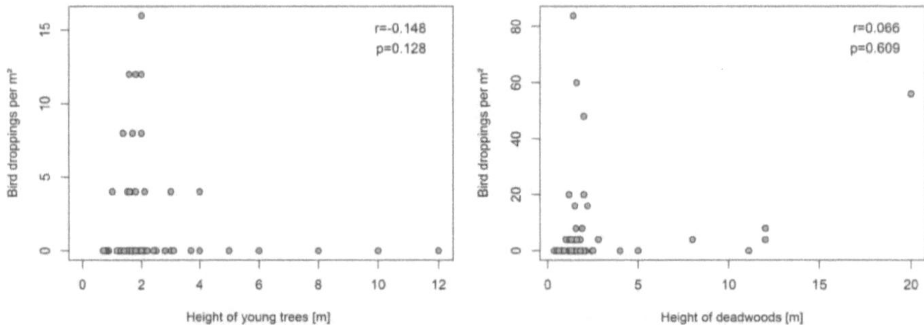

Fig. 4.6 Measured bird droppings [n m⁻²] under young trees (left) and deadwoods (right) on all open areas plotted against the heights [m] of the structural elements (p - *p*-value of correlation, r - Spearman correlation coefficient).

4.5 Discussion

The recorded densities of frugivorous bird droppings on traps can be used as an indicator of the birds' preferred resting places in disturbed forest areas, for example, areas damaged by storm.

The results of the study revealed that structural elements are a requirement for the input of seed by means of endozoochorous seed dispersal by frugivorous birds, as droppings were recorded exclusively in the vicinity of structural elements. Storm-felled areas without structural elements have low potential for sufficient natural regeneration of tree species by means of endozoochorous seed dispersal by birds (see McDonnell & Stiles 1983, Jordano & Schupp 2000). Comparable results were documented by Ferguson and Drake (1999) and Żywiec and Ledwoń (2008) who found bird droppings mostly under trees of closed forests, and rarely in large gaps or open areas without structural complexity. McDonnell and Stiles (1983), Kollmann (2000) and Żywiec and Ledwoń (2008) mentioned that birds prefer perches or resting places that protect them against predators and offer a wide field of view during defecation. Therefore, structural elements in open areas, which can be used as perches, attract birds and increase seed input through bird droppings (McDonnell & Stiles 1983, Ferguson & Drake 1999, Kollmann 2000).

Jordano and Schupp (2000), Stiebel (2003), Żywiec et al. (2013) and Żywiec (2014) mentioned that birds usually defecate seeds within a radius of 30-40 m around the seed source due to the short retention time of the indigestible seeds in the small guts of frugivorous birds of 30 minutes to a maximum of two hours (Barnea et al. 1992, Bonn & Poschlod 1998). This results in so-called seed shadows with high seed densities in the immediate vicinity of seed trees and resting places (Stimm & Böswald 1994, Jordano & Schupp 2000). This behaviour of frugivorous birds, their movements and their species-specific habitat preferences influence the spatial patterns of seed distribution significantly (Hoppes 1987, Jordano & Schupp 2000, Nathan & Muller-Landau 2000). To facilitate the arrival of seed of endozoochorously dispersed tree species on large disturbed forest sites, a complete clearing of 'key structures' on storm-felled areas should be avoided.

The results of this study showed a preference of frugivorous birds for standing structural elements and elements at least 1 m high. This was reflected in the findings presented by McDonnell and Stiles (1983) and McDonnell (1986). Those authors mentioned that structural elements with prevailing vertical structures, horizontal branches and a sufficient perch height of at least 1.5 m are the resting places most frequently chosen by birds on open areas, independent of crown cover and the resulting predator protection. In our study, bird droppings were only found in the vicinity of young trees where these did not exceed 4 m in height. We assume this upper limit may be a methodical restriction associated with the positioning of the dropping traps on the ground. With increasing tree height and increasing branchiness, the probability of faeces being caught on the dropping traps decreases. We assume that droppings were probably intercepted by one of the many branches above the dropping trap, as droppings were recorded on traps under deadwood structures up to 20 m in height.

The results of this study also revealed that birds rest on dead tree parts more often than on other structural elements. Young birch, rowan and spruce trees of low dimensions were avoided by frugivorous birds, probably due to the vertically-oriented and flexible branches. The highly elastic thin twigs and branches and their propensity to bend under the birds' weight may have rendered them unattractive as resting places. The horizontally-oriented, less flexible branches of European beech trees were more frequently used as perches but the numbers of young beech trees on these higher elevation sites in the Thuringian forest is low, especially after disturbances. Artificial structural elements without horizontal structures, such as fences or tree protectors, appeared not to be viable alternatives to deadwood structures or young trees (see Stimm & Böswald 1994).

4.6 Conclusions for silvicultural practice

The best conditions to enhance endozoochorous seed dispersal by frugivorous birds in storm-felled areas can be created by (i) leaving lying and standing deadwood with larger branches, (ii) leaving a sufficient number of tall stumps with branches and (iii) leaving upturned root plates. These elements promote post-foraging seed input by frugivorous birds and accelerate succession (see Stimm & Böswald 1994, Dale et al. 2001, Stiebel 2003). Structural elements are also useful for certain bird species as breeding and nesting places (Hunter 1999, Fuller 2013).

Several studies investigating ornithology and regeneration in connection with endozoochorous seed dispersal have shown that birds are able to move seeds from source trees over distances of 100 m (Leder 1992, Bakker et al. 1996, Stiebel 2003). Where already established, young trees of species dispersed by birds should be maintained on storm-felled areas, even though they are not often used by birds as resting places. Fruit-bearing tree species, such as rowan, produce fruits at an early age and in abundant quantities in the absence of competition pressure (Gockel 2016). The young trees established on disturbed areas contribute to a spatial network of seed sources and make open areas more attractive for birds due to the availability of fruits (McDonnell & Stiles 1983, Herrera et al. 1994, Albrecht et al. 2012). This will further encourage endozoochorous seed flow into open areas.

In terms of the practical management of the regeneration of storm-felled areas or other disturbed forest areas it can be concluded that complete site clearance, connected with a reduction of structural diversity, is justified neither from an economic nor an ecological perspective. The only legitimate reason for partial site clearance might be the removal of operational hazards, which may originate from obstacles such as upturned root plates or standing deadwood. Retained seed trees of the desired species are a prerequisite to preserve the connectivity of species between storm-felled sites and neighbouring forests (Gregor & Seidling 1997, Gockel 2016).

4.7 References

Albrecht J, Neuschulz EL, Farwig N (2012). Impact of habitat structure and fruit abundance

on avian seed dispersal and fruit predation. Basic Applied Ecology Journal 13: 347-354. - doi: https://doi.org/http://dx.doi.org/10.1016/j.baae.2012.06.005

Argus GW (2006). Guide to *Salix* (willow) in the Canadian Maritime Provinces (New Bruns-

wick, Nova Scotia, and Prince Edward Island). Canadian Museum of Nature, Ottawa, Canada, pp. 49. [online] URL: http://accs.uaa.alaska.edu/files/botany/publications/2006/GuideSalixCanadianAtlanticMaritime.pdf

Bakker JP, Poschlod P, Strykstra RJ, Bekker RM, Thompson K (1996). Seed banks and seed dispersal: important topics in restoration ecology. Acta Botanica Neerlandica 45, 461-490. - doi: https://doi.org/10.1111/j.1438-8677.1996.tb00806.x

Barnea A, Yom-Tov Y, Friedmann J (1992). Effect of frugivorous birds on seed dispersal and germination of multi-seeded fruits. Acta Oecologica 13: 209-219. - doi: https://doi.org/10.1023/A:1013282313035

Baxter P, Jack S (2008). Qualitative case study methodology: study design and implementation for novice researchers. The Qualitative Report 13: 544-559. - doi: https://nsuworks.nova.edu/tqr/vol13/iss4/2.

Bonn S, Poschlod P (1998). Ausbreitungsbiologie der Pflanzen Mitteleuropas [Biology of plant distribution in central Europe]. Quelle & Meyer Verlag, Wiesbaden, Germany, pp. 404.

Burse K-D, Schramm H-J, Geiling S, Meinhardt H, Schölch M (1997). Die forstlichen Wuchsbezirke Thüringens - Kurzbeschreibung [The forest growth areas of Thuringia - short description]. Mitteilungen der Landesanstalt für Wald und Forstwirtschaft, Gotha, Germany, pp. 199.

Bushart M, Suck R (2008). Potenzielle natürliche Vegetation Thüringens [Possible natural vegetation of Thuringia]. Schriftenreihe der Thüringer Landesanstalt für Umwelt und Geologie, Jena, Germany, pp. 139.

Dale VH, Joyce LA, McNulty S, Neilson RP, Ayres MP, Flannigan MD, Hanson PJ, Irland LC, Lugo AE, Peterson CJ, Simberloff D, Swanson FJ, Stocks BJ, Wotton BM (2001). Climate change and forest disturbances. BioScience 51: 723-734. - doi: https://doi.org/10.1641/0006-3568(2001)051[0723:CCAFD]2.0.CO;2

Erlbeck R (1998). Die Vogelbeere (*Sorbus aucuparia*) - ein Porträt des Baumes des Jahres 1997 [The rowan (*Sorbus aucuparia*) - a portrait of the tree of the year 1997]. In: "Beiträge zur Vogelbeere [Contributions to the rowan]" (Schmidt O ed). LWF-Wissen 17, Germany, pp. 2-14.

Ferguson RN, Drake DR (1999). Influence of vegetation structure on spatial patterns of seed deposition by birds. New Zealand Journal of Botany 37: 671-677. - doi: https://doi.org/10.1080/0028825X.1999.9512661

Fink AH, Brücher T, Ermert V, Krüger A, Pinto JG (2009). The European storm Kyrill in January 2007: synoptic evolution, meteorological impacts and some considerations with respect to climate change. Natural Hazards and Earth System Sciences 9: 405-423. - doi: https://doi.org/10.5194/nhess-9-405-2009

Frischbier N, Profft I, Hagemann U (2014). Potential impacts of climate change on forest habitats in the Biosphere Reserve Vessertal-Thuringian Forest in Germany. In: "Managing Protected Areas in Central and Eastern Europe under Climate Change - Advances in Global Change Research" (Rannow S, Neubert M eds). Springer Science and Business Media, Dordrecht, Germany, pp. 243-257.

Fuller RJ (2013). Searching for biodiversity gains through woodfuel and forest management. Journal of Applied Ecology 50: 1295-1300. - doi: https://doi.org/10.1111/1365-2664.12152

García D, Martínez I, Obeso JR (2007). Seed transfer among bird-dispersed trees and its consequences for post-dispersal seed fate. Basic and Applied Ecology 8: 533-543. - doi: https://doi.org/doi:10.1016/j.baae.2006.11.002

Gauer J, Aldinger E (2005). Waldökologische Naturräume Deutschlands: Forstliche Wuchsgebiete und Wuchsbezirke - mit Karte 1:1.000.000 [Forest ecological natural areas of Germany: forest growth areas and growth districts - with map 1:1,000,000]. Mitteilungen des Vereins für Forstliche Standortskunde und Forstpflanzenzüchtung, Stuttgart, Germany, pp. 324.

Gockel S (2016). Wachstumsreaktionen einzeln eingemischter Vogelbeeren (*Sorbus aucuparia* L.) in Fichtenjungbeständen nach Freistellung [Growth reactions of single mixed rowans (*Sorbus aucuparia* L.) in young spruce stands after thinning]. PhD book, Faculty of Environmental Science, Dresden University of Technology, Dresden, Germany, pp. 348.

Gregor T, Seidling W (1997). 50 Jahre Vegetationsentwicklung auf einer Schlagfläche im osthessischen Bergland [50 years' succession on a woodland clearing in the mountain area of eastern Hesse]. Forstwissenschaftliches Centralblatt 116: 218-231. - doi: https://doi.org/10.1007/BF02766899

Guitian J, Munilla I (2010). Responses of mammal dispersers to fruit availability: Rowan (*Sorbus aucuparia*) and carnivores in mountain habitats of northern Spain. Acta Oecologica 36: 242-247. - doi: https://doi.org/10.1016/j.actao.2010.01.005

Hacker H (1999). Die Insektenwelt der Weiden [The insects on willows]. In: „Beiträge zur Silberweide [Contributions to the white willow]" (Schmidt O ed.), LWF 24, Germany, pp. 25-27.

Herrera CM, Jordano P, López-Soria L, Amat JA (1994). Recruitment of a mast-fruiting, bird-dispersed tree: Bridging frugivore activity and seedling establishment. Ecological Monographs 64: 315-344. - doi: https://doi.org/10.2307/2937165

Hoppes WG (1987). Pre- and post-foraging movements of frugivorous birds in an eastern deciduous forest woodland, USA. Oikos 49: 281-290. - doi: 10.2307/3565762

Hunter ML (1999). Maintaining biodiversity in forest ecosystems. Cambridge University Press, Cambridge, UK, pp. 716.

Huth F (2009). Untersuchungen zur Verjüngungsökologie der Sand-Birke (*Betula pendula* Roth) [Study of regeneration cycle of sand birch (*Betula pendula* Roth)]. PhD book, Faculty of Forest, Geo and Hydro, Dresden University of Technology, Dresden, Germany, pp. 383.

Hynynen J, Niemistö P, Viherä-Aarnio A, Brunner A, Hein S, Velling P (2010). Silviculture of birch (*Betula pendula* Roth and *Betula pubescens* Ehrh.) in northern Europe. Forestry 83: 103-119. - doi:10.1093/forestry/cpp035

Jordano P, Schupp EW (2000). Seed disperser effectiveness: The quantity component and patterns of seed rain for *Prunus mahaleb*. Ecological Monographs 70: 591-615. - doi: https://doi.org/10.1890/0012-9615(2000)070[0591:SDETQC]2.0.CO;2

Karlsson M (2001). Natural regeneration of broadleaved tree species in southern Sweden - Effects of silvicultural treatments and seed dispersal from surrounding stands. PhD book,

Southern Swedish Forest Research Centre, Swedish University of Agricultural Sciences, Alnarp, Sweden, pp. 44.

Kay QON (1985). Nectar from willow catkins as a food source for blue tits. Bird Study 32: 40-44. - doi: https://doi.org/10.1080/00063658509476853

Kollmann J (2000). Dispersal of fleshy-fruited species: a matter of spatial scale? Perspectives in Plant Ecology, Evolution and Systematics 3: 29-51. - doi: https://doi.org/10.1078/1433-8319-00003

Kuzovkina YA, Quigley MF (2005). Willows beyond wetlands: uses of *Salix* L. species for environmental projects. Water, Air, & Soil Pollution 162: 183-204. - doi: https://doi.org/https://doi.org/10.1007/s11270-005-6272-5

Leder B (1992). Weichlaubhölzer: Verjüngungsökologie, Jugendwachstum und Bedeutung in Jungbeständen der Hauptbaumarten Buche und Eiche [Pioneer tree species: Regeneration, youth growth and importance in young stands of the main tree species beech and oak]. Landesanstalt für Forstwirtschaft Nordrhein-Westfalen, Arnsberg, Germany, pp. 416.

Löf M, Bergquist J, Brunet J, Karlsson M, Welander NT (2010). Conversion of Norway spruce stands to broadleaved woodland – regeneration systems, fencing and performance of planted seedlings. Ecological Bulletins 53: 165-173.

McDonnell MJ (1986). Old field vegetation height and the dispersal pattern of bird-disseminated woody plants. Bulletin of the Torrey Botanical Club 113: 6-11. - doi: https://doi.org/10.2307/2996227

McDonnell MJ, Stiles EW (1983). The structural complexity of old field vegetation and the recruitment of bird-dispersed plant species. Oecologia 56: 109-116. - doi: 10.1007/BF00378225

McVean DN (1956). Ecology of *Alnus glutinosa* (L.) Gaertn.: VI. Post-glacial history. Journal of Ecology 44: 331-333. - doi: 10.2307/2256825

Nathan R, Muller-Landau HC (2000). Spatial patterns of seed dispersal, their determinants and consequences for recruitment. Tree 15: 278-285. - doi: https://doi.org/https://doi.org/10.1016/S0169-5347(00)01874-7

Obeso JR, Martínez I, García D (2011). Seed size is heterogeneously distributed among destination habitats in animal dispersed plants. Basic and Applied Ecology 12: 134-140. - doi: https://doi.org/10.1016/j.baae.2011.01.003

Paulsen TR, Högstedt G (2002). Passage through bird guts increases germination rate and seedling growth in *Sorbus aucuparia*. Functional Ecology 16: 608-616. - doi: https://doi.org/10.1046/j.1365-2435.2002.00668.x

Perala DA, Alm AA (1990). Reproductive ecology of birch: a review. Forest Ecology and Management 32: 1-38. - doi:10.1016/0378-1127(90)90104-J

Profft I (2013): Waldumbau in den mittleren, Hoch- und Kammlagen des Thüringer Waldes [Forest restoration in the middle, high elevations and along the ridges of the Thuringian Forest]. Auftaktkolloquium - Waldumbau in den mittleren, Hoch- und Kammlagen des Thüringer Waldes. ThüringenForst-Anstalt öffentlichen Rechts, Gotha, Germany, 28.02.2013.

Raspé O, Findlay C, Jacquemart A-L (2000). *Sorbus aucuparia* L. Journal of Ecology 88: 910-930. - doi: 10.1046/j.1365-2745.2000.00502.x

R Core Team (2014). R: A Language and Environment for Statistical Computing. R Foundation for Statistical Computing, Vienna. Web Site. URL: https://www.R-project.org

Schmidt O (1999). Vogelwelt und Weiden [Birds and willows]. In: „Beiträge zur Silberweide [Contributions to the white willow]" (Schmidt O ed.), LWF 24, Germany, pp. 21-24.

Schmidt O (1998). Vogelbeere und Tierwelt [Rowan and animals]. In: "Beiträge zur Vogelbeere [Contributions to the rowan]" (Schmidt O ed). LWF-Wissen 17, Germany, pp. 59-64.

Stiebel H (2003). Frugivorie bei mitteleuropäischen Vögeln [Frugivorie of central European birds]. PhD book, Faculty of mathematics and natural sciences, Carl-von-Ossietzki-University of Oldenburg, Wilhelmshaven, Germany, pp. 219.

Stimm B, Böswald K (1994). Die Häher im Visier. Zur Ökologie und waldbaulichen Bedeutung der Samenausbreitung durch Vögel [The jays in the sights. The ecology and silvicultural importance of seed dispersal by birds]. Forstwissenschaftliches Centralblatt 113: 204-223. - doi: https://doi.org/10.1007/BF02936698

Waesch, G., 2003. Montane Graslandvegetationen des Thüringer Waldes: Aktueller Zustand, historische Analyse und Entwicklungsmöglichkeiten [Montane grassland vegetation of the Thuringian Forest: current state, historical analysis and development opportunities]. Cuvillier Verlag, Göttingen, Germany, pp. 219.

Wagner S, Wälder K, Ribbens E, Zeibig A (2004) Directionality in fruit dispersal models for anemochorous forest trees. Ecological Modelling 179: 487-498. - doi: https://doi.org/https://doi.org/10.1016/j.ecolmodel.2004.02.020

Worrell R (1995). European aspen (*Populus tremula* L.): a review with particular reference to Scotland: I. Distribution, ecology and genetic variation. Forestry 68: 93-105. - doi: 10.1093/forestry/68.2.93

Zar JH (2010). Biostatistical analysis. Prentice Hall, Upper Saddle River, New Jersey, USA, pp. 944.

Zerbe S (2009). Renaturierung von Waldökosystemen [Restoration of forest ecosystems]. In: "Renaturierung von Ökosystemen in Mitteleuropa [Restoration ofecosystemsin central Europe]" (Zerbe S, Wiegleb G eds.). Spektrum Akademischer Verlag, Heidelberg, Germany, pp. 153-182.

Żywiec M (2014). Seedling survival under conspecific and heterospecific trees: the initial stages of regeneration of *Sorbus aucuparia*, a temperate fleshy-fruit pioneer tree. Annales Botanici Fennici 50: 361-371. - doi: https://doi.org/10.5735/085.050.0611

Żywiec M, Holeksa J, Wesołowska M, SzewczykJ, Zwijacz-Kozica T, Kapusta P (2013). *Sorbus aucuparia* regeneration in a coarse-grained spruce forest - a landscape scale. Journal of Vegetation Science 24: 735-743. - doi:10.1111/j.1654-1103.2012.01493.x

Żywiec M, Ledwoń M (2008). Spatial and temporal patterns of rowan (*Sorbus aucuparia* L.) regeneration in West Carpathian subalpine spruce forest. Plant Ecology 194: 283-291. - doi: 10.1007/s11258-007-9291-z

Chapter 5

Soil seed banks of pioneer tree species in European temperate forests: a review

Katharina Tiebel[1], Franka Huth[1], Sven Wagner[1] (2018)

iForest 11: 48-57. - doi: 10.3832/ifor2400-011

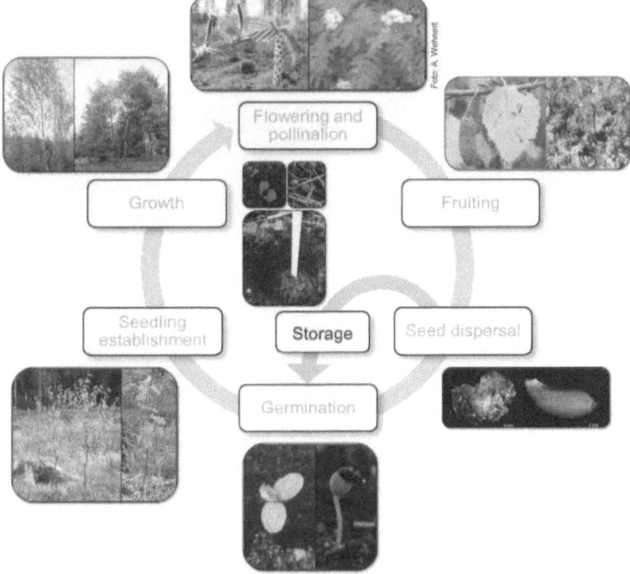

[1]TU Dresden, Institute of Silviculture and Forest Protection, Chair of Silviculture, Pienner Str. 8, 01737 Tharandt, Germany

5.2 Introduction

Soil seed banks are an important component in the succession and regeneration of ecosys-tems. Soil seed banks are buried seed reserves, which are viable and able to germinate under changing environmental conditions (Fenner 1985, Thompson et al. 1997, Berger et al. 2004). The formation of a soil seed bank is a strategy developed by plants to prevent germination under unfavorable soil and climate conditions (Bradbeer 1988, Leck et al. 2008, Saatkamp et al. 2014). In disturbed areas of forest, seeds of different species are granted an opportunity to germinate and cover the open soil surface, even though these species may not have been represented in this area for long time (Fenner 1985, Bossuyt & Hermy 2001). Soil seed banks could contribute significantly to the reforestation of disturbed woodlands. They may also compensate for a recent absence of seed sources within or around a damaged area.

Soil seed banks of forests generally exhibit lower species diversity and seed densities than those present in other ecosystems (Kalamees & Zobel 1998, Hopfensperger 2007, Bossuyt & Honnay 2008). Deciduous, young or managed forests are characterized by larger seed numbers and greater species richness than coniferous, older or unmanaged forests (Donelan & Thompson 1980, Bossuyt & Hermy 2001, Godefroid et al. 2006, Ebrecht & Schmidt 2008, Plue et al. 2010). The seed bank compositions of northern and western European forests differ from those of eastern European forests (Bossuyt et al. 2002). The composition of seed banks and ground flora in forests also differ from each other (Bossuyt & Hermy 2001, Bossuyt et al. 2002, Zobel et al. 2007). In central European temperate forests, soil seed banks predominantly contain herbaceous plant species of early or middle successional stages. The seed banks are refreshed by seeds of species that emerge in case of disturbance in forest ecosystems. Species of early or middle successional stages are light demanding species, adapted to disturbances, and able to form a persistent soil seed bank (Donelan & Thompson 1980, Bossuyt et al. 2002, Godefroid et al. 2006). Hopfensperger (2007) suggested that pioneer species, present in early successional stages, can form a persistent seed bank at the beginning of succession to woodland. Seeds of ancient, shade-tolerant forest species, shrubs and tree species in general, are not well represented in the soil seed bank, because the seeds of these species do not remain viable for long (Donelan & Thompson 1980, Bossuyt & Hermy 2001). However, pioneer tree species are also regarded as light demanding species. In Europe, *Betula* spp., *Salix* spp., *Populus* spp., *Alnus* spp. and *Sorbus aucuparia* L. represent deciduous pioneer tree species. These tree species are short-lived species, which produce large quantities of seeds, have long seed dispersal distances and exhibit fast juvenile growth (Perala & Alm 1990, Raspé et al. 2000, Zerbe 2001). Pioneer tree species are very common in early successional stages and in disturbed woodlands in central Europe (Zerbe 2001). With climate change, and the associated increase in the frequency and intensity of disturbances (e.g., storm events – Seidl et al. 2014), pioneer tree species are of growing importance for natural reforestation, and consequently their soil seed banks too. Pioneer tree species can regenerate rapidly and successfully colonize large areas in years of high seed production (Perala & Alm 1990, Leder 1992, Raspé et al. 2000, Argus 2006). As a consequence, pioneer tree species can mitigate negative consequences associated with disturbed areas, such as soil erosion and the loss of nutrients (Barnes et al.

1998, Schölch 1998, Argus 2006, Zerbe 2009, Fischer et al. 2016). However, pioneer tree species exhibit irregular seed production patterns (mast years – Sarvas 1952, Bjorkbom 1967, Holm 1994, Osumi & Sakurai 1997, Sperens 1997, Hynynen et al. 2010). A question that arises is whether pioneer tree species have the potential to regenerate from a soil seed bank in non-mast years, as shown by Hopfensperger (2007) for pioneer species. Currently little is known about the capacity of pioneer tree species in European temperate forests to establish seed banks, or how long their seeds persist in soil. Some burial experiments showed that rowan and birch seeds remain viable in soil for more than 5 years (Miles 1974, Granström & Fries 1985, Granström 1987, Skoglund & Verwijst 1989) and sometimes viable birch and willow seeds were detected in soil samples collected from deeper mineral soil layers (Hill & Stevens 1981, Staaf et al. 1987, Bakker et al. 1996a, Kalamees & Zobel 1998, Dölle & Schmidt 2009). However, the ability of pioneer tree species to form at least a short-term seed bank is controversial in the literature. The short viability period of pioneer tree seeds after dispersal is often mentioned and many authors support the hypothesis that pioneer tree species do not generally form a seed bank (Hill & Stevens 1981, Amezaga & Onaindia 1997, Buckley et al. 1997, Ebrecht & Schmidt 2008, Heinrichs 2010). By contrast, Granström & Fries (1985), Osumi & Sakurai (1997), Erlbeck (1998), Rydgren et al. (1998) and Decocq et al. (2004) suggested that birch, alder and rowan may make up part of the forest seed bank. If pioneer tree species have the capacity to establish a seed bank, years with low levels of fructification can be compensated for and the colonization of open areas, for example, would not depend exclusively on annual seed rain (Osumi & Sakurai 1997).

In this review, available data pertaining to densities of birch, alder, willow, poplar and rowan in soil seed banks in central and north-west European temperate forests are documented based on a survey of the literature. The aim was to summarize the general findings and to identify knowledge gaps concerning the soil seed bank with respect to these short-lived tree species. This species-specific information were discussed in the context of the meaning of the soil seed bank, and in relation to disturbance regimes, succession and reproductive ecology.

5.3 Methods of literature search

Our review is based on studies carried out in central and north-west European temperate forests published in the period 1979-2013. The keywords "seed bank", "propagule bank" and "buried seeds" were used in combination with either "forest" or "woodland". An article was selected when the seed density per m² could be calculated in order to make the results comparable with those of other studies. A total of 33 studies from 14 countries matched the criteria

(Fig. 5.1). Most of papers were found by searching the "Web of Science" database, meaning the papers had to be published in international peer reviewed journals with an impact factor. Only 3 papers included in the review were published in non-peer reviewed journals, 2 of which were written in English (Ebrecht & Schmidt 2008, Heinrichs 2010, Jedrzejczak 2013). These papers were found through citations within other international soil seed bank papers. The forests presented in all of the chosen studies were considered to be distinct sample plots wherever the authors classified the study sites and their sample plots as independent (e.g., Staaf et al. 1987, Dougall & Dodd 1997, Dölle & Schmidt 2009). In this way, 136 sample plots were recorded, which differed in their histories, forest types, stand ages, canopy densities and management strategies (see Table S1 in Supplementary material, p 131). The mean seed density per m² of birch (*Betula* spp.), alder (*Alnus glutinosa* [L.] Gaertn.), aspen (*Populus tremula* L.) and willow (*Salix* spp.) was calculated for each plot. The soil samples differed in their depths and in terms of the soil layers. Authors took samples from humus and mineral soil, or only from the mineral soil layer. In some cases no information about whether litter and humus were removed prior to sampling was provided (Table 5.1). In this paper the term "birch" is used to represent *Betula pendula* and *B. pubescens*, with "*Salix* spp." used to indicate all willow species detected in soils. This corresponds to the approaches used by the authors of the selected studies.

Fig. 5.1 Location of the 33 selected soil seed bank studies in Europe.

Table 5.1 Summary of the 33 selected seed bank studies in central and north-west European temperate forests and the information about seed densities provided. (‡) Depth of core: (?) not clear whether the humus layer was tested or not; (-) humus layer not tested; (+) humus layer tested separately; (++) humus and mineral soil layer tested together; (5) soil sample depth of 0-5 cm in the mineral soil. (†) Temperature: (1) cold stratification of soil samples before seedling-emergence treatment; (2) cold stratification of soil samples integrated within the seedling-emergence treatment. (§) Species: (?) the species or genus with individually defined small numbers in the soil was excluded from the presentation; (+) species or genus detected; (-) species or genus not detected; (-/-) tree species excluded, which germinated in sterile control trays.

Author(s) & Year	Date of soil sampling	Depth of core [‡] (cm)	Temperature [†] (range °C or a day/night)	Artificial light (day/ night)	Seed extraction method	Study duration (months)	*Betula* ssp.	*Alnus glutinosa*	*Salix* ssp.	*Populus tremula*	*Sorbus aucuparia*
Amrein et al. 2005	Jan	(-) 10	16 °C	16/8 h		5	-	-	-	-	-
Augusto et al. 2001	Feb	(-) 5	nursery and closed shade house	-		7	+	-	-/-	-	-
Bakker et al. 1996a	Apr	(-) 10	25 - 45/15 °C		+	4.5	+	-	+	-	-
Bekker et al. 2000	Mar	(?) 10	25/15 °C	12/12 h	+	3	+	+	+	-	-
Berger et al. 2004	Feb, Mar, Jun, Sept	(+) 35	glasshouse	6-9 pm		3-4	+	-	+	-	-
Bossuyt et al. 2002	Mar, Sept	(-) 20	14 - 25 °C	16/8 h		5	+	-	-	-	-
Brown & Oosterhuis 1981	Apr	(-) 15	glasshouse [1]	-		24	+	-	+	-	-
Buckley et al. 1997	Nov-Feb	(?) 10	unheated polythene tunnel	-		6-9	+	-	-	-	-
Decocq et al. 2004	Jun	(-) 20	20/16 °C	12/12 h		6 [2]	+	+	-	-	-
Dölle & Schmidt 2009	Mar	(-) 30	unheated glasshouse			12	+	-	+	-	-
Donelan & Thompson 1980	May	(?) 7	unheated glasshouse			3	+	-	-	-	-
Dougall & Dodd 1997	Apr	(-) 10	polythene tunnel/glasshouse			4	+	-	-	-	-
Ebrecht & Schmidt 2008	Mar	(+) 10	unheated glasshouse			10	+	-	-	-	-
Falińska 1999	Mar/Apr, Sep/ Oct	(?) 3	18 - 22 °C		+	36	+	+	+	-	-
Grandin 2001	Jul	(+) 15	22/5 °C [1]	16/8 h		16 [2]	+	?	?	?	?
Granström 1982	Jul	(+) 5	22/12 °C [1]	18/6 h	+	6	+	-	-	-	-
Granström 1988	Apr	(+) 6	22/12 °C [1]	18/6 h		63 [2]	+	-	+	-	-
Heinrichs 2010	Mar	(+) 20	unheated glasshouse			12	+	?	-/-	+	?
Hester et al. 1991	May	(++) 5	glasshouse			12	+	-	-	-	-
Hill & Stevens 1981	Apr	(+) 10	unheated glasshouse			8	+	-	-	-	-
Jankowska-Błaszczuk 1998	Mar	(?) 5	unheated glasshouse			43	+	-	-	-	-
Jankowska-Błaszczuk et al. 1998	early spring	(-) 10	unheated glasshouse			8	+	-	-	-	-
Jaroszewicz 2013	Jun	(-) 10	unheated glasshouse			25	+	?	?	?	?
Jędrzejczak 2013	Oct, Nov	(?) 10	18 - 24 °C	12/12 h		5	+	-	-	-	-
Kalamees & Zobel 1998	May	(-) 10	unheated glasshouse			> 4	+	-	-	-	-
Kjellsson 1992	Mar, Apr	(++) 17.5	22/12 °C [1]	16/8 h		4	+	+	-	-	-
Komulainen et al. 1994	Jun	(++) 10	25 °C		+	> 10 [2]	+	-	-	-	-
Milberg 1995	May	(?) 8	20/8 °C [1]	16/8 h		12 [2]	+	?	?	?	?
Miller & Cummins 2003	Jul, Aug	(++) 5	6 - 25 °C [1]	12/12 h		12.5	+	-	-	-	-
Mitschell et al. 1998	Feb	(?) 6.3	polyethylene tunnel			15 [2]	+	-	-	-	-
Staaf et al. 1987	Apr	(+) 5	unheated glasshouse (10 - 30 °C)			3	+	-	+	-	-
Thompson & Grime 1979	Oct-Oct (every 6 weeks)	(++) 3	20/15 °C	16/8 h		1.25	-	-	-	-	-
Warr et al. 1994	May, Jun	(?) 15	shade tunnel			10-12	+	-	-	-	-

5.4 Species-specific reproductive ecology determining the potential of soil seed banks

It is often assumed by practitioners that a bountiful fructification of pioneer tree species recurs annually. However, like intermediate and climax tree species, short-lived species exhibit irregular seed production patterns, influenced by soil and climate conditions, and the individual fitness of seed trees (Sarvas 1952, Bjorkbom 1967, Holm 1994, Osumi & Sakurai 1997, Sperens 1997, Hynynen et al. 2010). The germination percentage of the seeds also varies from year to year, with mast years usually characterized by the highest germination rates (Sarvas 1952, Bjorkbom 1967, Holm 1994, Osumi & Sakurai 1997, Sperens 1997, Raspé et al. 2000, Hynynen et al. 2010). However, pioneer tree species exhibit seed morphologies, seed dispersal distances as well as requirements for germination and seedling establishment that are different from those of intermediate and late-successional species (McVean 1953, Atkinson 1992, Lautenschlager 1994, Worrell 1995, Raspé et al. 2000). Despite differences between their fruits and seeds, birch and alder (winged nuts), willow and poplar (catkins, seeds with pappus) and rowan (small seeds within a red fleshy fruit) can be analyzed together as each group possesses morphological similarities (McVean 1956, Perala & Alm 1990, Worrell 1995, Raspé et al. 2000).

Birches can produce 2-10 million winged seeds per tree (Perala & Alm 1990, Huth 2009), which are 1.5-2.0 mm in size without the wings (Brouwer & Stählin 1975). Seed rain takes place mainly from June to November (Perala & Alm 1990, Huth 2009). From November until the end of the following June, the seed rain falls to less than 100 seeds per m² (Huth 2009). Mean dispersal distances by wind vary between 60 and 100 m (Fries 1984 cited in Perala & Alm 1990, Karlsson 2001, Huth 2009), but the highest seed densities are deposited within distances of 25-50 m (Bjorkbom 1971, Fries 1984 cited in Perala & Alm 1990 - Fig. 5.2). Maximum propagation distances amount to 550-800 m (Huth 2009). Most seeds germinate in spring after dispersal (Perala & Alm 1990). Alder seed trees generally produce lower seed numbers (240,000 seeds per tree) than birch. The diaspores of alder have smaller wings and larger seed nuts (2.0-2.5 mm), and their mean and maximum dispersal distances from seed trees are 30 m and 60 m, respectively (McVean 1953, Brouwer & Stählin 1975). High seed densities were found within distances of less than 10 m (McVean 1953, 1956, Karlsson 2001). Seeds of alder trees are mature in November, but most are only released in February and March, and, like birch, germinate predominantly after dispersal in spring (Pietzarka & Roloff 2010). Seeds of birch and alder do not exhibit dormancy (McVean 1953, Atkinson 1992). The spatial distribution of deposited seeds on soil depends on the position of the seed trees and on the dispersal agents (Bakker et al. 1996b). The distribution of wind-dispersed seeds tends to be non-random (Greene & Johnson 1996). This applies especially to birch and alder but also to willow and poplar. Deposited seeds of these species are often spatially aggregated because the seeds are brought in dense infructescenses (Hill & Stevens 1981, Kjellsson 1992, Dougall & Dodd 1997).

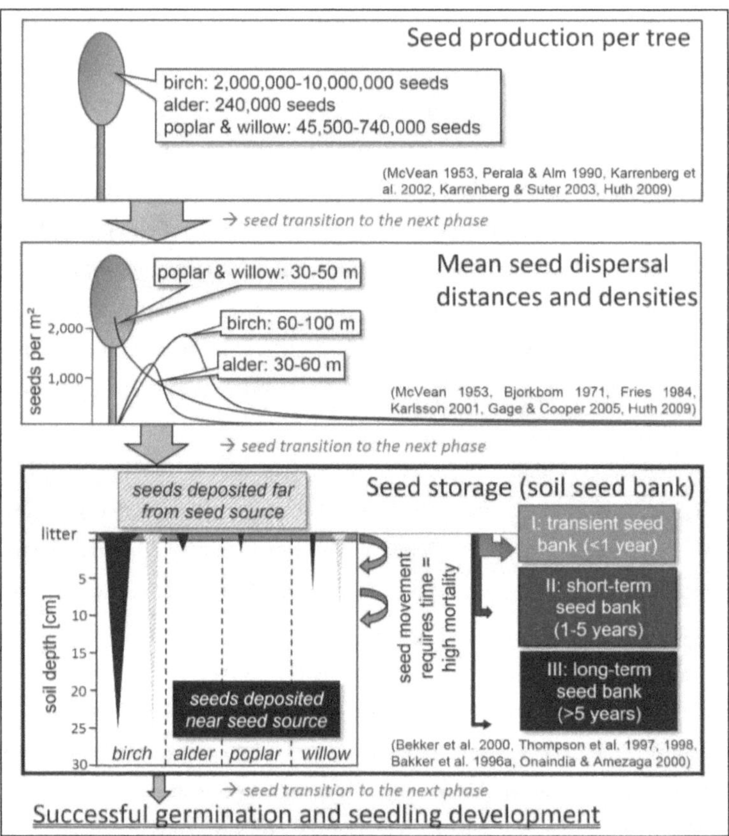

Fig. 5.2 Schematic diagram showing the relation between seed production, seed dispersal and seed storage in soil of birch, alder, poplar and willow.

Willows and poplars can produce between 45,500 and 740,000 seeds per tree (Karrenberg et al. 2002, Karrenberg & Suter 2003). The wind-dispersed (anemochorous) willow and poplar seeds possess hairs to facilitate flight and range in size between 0.8-1.5 mm and 1.0-2.5 mm, respectively (Brouwer & Stählin 1975). Maximum willow seed dispersal distances of 2-3 km are much longer than for birch and alder (Schirmer 2006). For this reason, Gage & Cooper

(2005) always adopted a background level (i.e., "noise") of 10-30 seeds per m². Nevertheless, the largest densities of deposited seeds were measured at distances of between 30-50 m from willow seed trees (Gage & Cooper 2005 - Fig. 5.2). The small seeds of both genera are shortlived (Worrell 1995, Barsoum 2002, Young & Clements 2003). Regarding early flowering (spring) poplars and willow species, like *Salix caprea* L., seed rain takes place from April to June (Chmelar & Meusel 1986). Seeds lying on soil that do not germinate immediately after dispersal lose their viability after 1-6 weeks (Junttila 1976, Niiyama 1990, Leder 1992, Worrell 1995, Karrenberg & Suter 2003). In contrast, seeds of late flowering (summer and autumn) willow species remain viable until the next spring (Chmelar & Meusel 1986), which means a lifespan of about half a year.

The fruits of rowan ripen between August and October (Raspé et al. 2000). Fruits and seeds are dispersed endozoochorously, by birds and small mammals. The seeds of rowan exhibit embryo and seed coat dormancy. If rowan fruits are not eaten by animals, the seeds germinate in the second year after maturation, when embryo and seed coat dormancy is broken under natural conditions (Raspé et al. 2000). Sometimes the seeds undergo a second period of dormancy when temperatures rise above 10 °C after winter or cold storage (Spethmann 2000). In such cases, seeds can remain viable for up to 5 years in the soil (Erlbeck 1998). Birds, as the main consumer of rowan fruits, have a significant influence on the spatial patterns of seed distribution. Bird droppings were mostly found under trees in more or less closed forests, and only very rarely in large gaps or open areas lacking structural complexity (Żywiec & Ledwoń 2008). Structural elements are used as perches by birds (McDonnell & Stiles 1983). Most frequently, birds drop seeds up to 40 m from seed trees (Żywiec et al. 2013). In the case of endozoochory, the density patterns of rowan seeds in the soil are clumped rather than randomly distributed (Clark et al. 1998, Żywiec et al. 2013).

5.5 Characterization and classification of soil seed banks

Simpson et al. (1989) emphasized that, *"all viable seeds present on or in soil or associated litter constitute the soil seed bank."* Seeds buried in the upper soil layer (litter and humus layer) have in most cases only been part of the seed bank for a short time. These seeds are probably part of the last seed rain, thus the character of the seed bank in the upper soil is prevailingly transient (Graber & Thompson 1978, Bakker et al. 1996a, 1996b, Osumi & Sakurai 1997, Houle 1998, Heinrichs 2010). Given the species-specific timing of seed rain, the time of sampling represents an important piece of information for the interpretation of seed densities in transient seed banks. Seeds found in deeper mineral soil layers are older and have persisted in the soil for a longer period (Kjellsson 1992, Bakker et al. 1996b, Thompson et al. 1997). Therefore, information on seed depth can be used as an indirect method to determine the seed longevity in the soil of a particular tree species. Buried seed experiments have contributed to a better understanding of the duration of seed viability in deeper soil layers. The experiments demonstrated what might happen to seeds in relation to viability, decomposition and mortality over time during storage in soil (Granström 1987, Skoglund & Verwijst 1989).

Litter cover and litter thickness are also two very important factors. Thick litter protects deposited seeds against movement, drought, predation and early germination, and so may help to maintain higher viability (Granström & Fries 1985, Egawa & Tsuyuzaki 2013). Almost no viable seeds were found on sites without litter protection. Thin litter, including seeds, for example, will be carried away by wind and accumulated at other places (Egawa & Tsuyuzaki 2013). Independent of the species, the number of viable seeds on and in peatland increased with litter thickness. The thickness of the humus layer determines the period of time required by seeds to penetrate litter and humus. During this time seeds are subject to mortality (Sarvas 1952, Van Tooren 1988, Holm 1994). Small, light and dry seeds without pulp are less prone to predation than larger seeds (Leck et al. 1989). Findings of glass bead experiments (the size and weight of the beads corresponded to rowan fruits) revealed that 30-40 % of all glass beads moved 1 cm in the soil over a 6-month period, whereas 4 % were transported 2-5 cm (Van Tooren 1988, Burmeier et al. 2010). Small seeds pass through litter faster, reaching the humus and mineral soil layer in a shorter time than larger seeds or beads (Egawa & Tsuyuzaki 2013). This is an advantage for pioneer tree species with regard to their strategy of fast colonization of disturbed areas: the small, light and dry seeds without pulp (Leck et al. 1989) achieve quicker contact with the mineral soil and can germinate successfully.

Soil seed banks can generally be classified according to the seed longevity of the species; that is, the period of time for which a seed stays viable and capable of germination. The classification most widely applied is that described by Thompson et al. (1997, 1998), who differentiated four types. The first type (I) includes all species with transient seed banks with a persistence in the soil of less than 1 year. Species with short-term persistent seeds (1-5 years) are assigned to type II. Type III is as a long-term persistent seed bank and includes species with seeds that persist in the soil for at least 4-5 years. Species that cannot be assigned to any seed bank type are combined in type IV.

5.5.1 Soil seed bank of *Betula* spp.

Houle (1998) found that birch seeds are viable for less than 2-3 years under field conditions, due to multiple causes of mortality. Mortality rose to 99 % under certain climate conditions (Houle 1998). However, 50-80 % of birch seeds buried artificially in soil were still viable after 3-5 years of storage (Granström 1987, Skoglund & Verwijst 1989). After 5 years stored birch seeds were partly degraded, but 50-60 % remained viable (Granström 1987). It is assumed that birch seeds decompose within a period of 5-7 years (Sarvas 1952). Skoglund & Verwijst (1989) concluded that, at a depth of 10 cm, birch seeds buried in the soil have a theoretical half-life in forest soil of roughly 13 years, but less than 1 year in the soil of wet meadows. In wet soils, early germination and fungal attacks lead to higher mortality rates (Harper 1955 cited in Ludwig et al. 1957). Birch seeds sown on bare ground germinated in the first year (Miles 1974). Therefore, seeds found in deeper mineral soil probably did not originate from the previous seed rain and so were part of the soil seed bank.

In most of the publications analyzed, birch was the only tree species exhibiting a high degree of consistency in soil seed banks in central and north-west European temperate forests (83 % of sample plots). Often birch was the second most abundant species of all, including herbaceous plant species. However, some authors found the extent of birch seed in the soil to be negligible (Hill & Stevens 1981, Staaf et al. 1987, Buckley et al. 1997). *Betula* species were present in all kinds of forest type, but seed densities depended on the presence or absence of seed sources. Large numbers of seeds can be found in the soil in the vicinity of seed trees (Bossuyt & Hermy 2001 - Fig. 5.2). Birch seed density ranged from 1 to 1100 seeds per m^2 in coniferous stands, mostly spruce and pine forests (Granström 1982, Komulainen et al. 1994, Warr et al. 1994, Dougall & Dodd 1997, Augusto et al. 2001, Miller & Cummins 2003, Berger et al. 2004, Heinrichs 2010, Jaroszewicz 2013). In deciduous forests 7-3850 viable seeds per m^2 were detected (Staaf et al. 1987, Kjellsson 1992, Warr et al. 1994, Dougall & Dodd 1997, Jankowska-Blaszczuk et al. 1998, Augusto et al. 2001, Bossuyt et al. 2002, Miller & Cummins 2003, Decocq et al. 2004, Jedrzejczak 2013). The seed density on succession sites ranged from 6 seeds per m^2 in a 4-year old Norway spruce clear cut (Heinrichs 2010) to a maximum of 3120 seeds per m^2 in a long-term overgrown grassland (Kalamees & Zobel 1998). Highest densities of 70-3760 seeds per m^2 were found in pure or birch-dominated stands (Hester et al. 1991, Kjellsson 1992, Warr et al. 1994, Jankowska-Blaszczuk 1998, Kalamees & Zobel 1998, Mitschell et al. 1998, Falińska 1999, Miller & Cummins 2003, Dölle & Schmidt 2009). However, 0 to 144 viable seeds per m^2 were detected in soils of deciduous and coniferous forests without any mature trees or seedlings in the proximity (Granström 1982,

Staaf et al. 1987, Amrein et al 2005). Hill & Stevens (1981) detected more viable birch seeds in a 4-year-old clear cut of a former Douglas fir plantation than seedlings in the vegetation layer, an indication that birch seeds remain viable in the soil for a longer time than is frequently assumed.

The birch seed density in different soil layers was reported for only a few samples. Various authors mentioned that birch seeds are mostly found in the humus or uppermost soil layers (Hill 1979, Granström 1988, Houle 1998, Sullivan & Ellison 2006). Houle (1998) concluded that less than 2 % of birch seed rain reaches the persistent seed bank. The numbers of birch seeds found at different soil depths ranged from 1-188 seeds per m² in the litter and humus layer (Granström 1982, Staaf et al. 1987) to 1-80 seeds per m² at a depth of 0-5 cm in mineral soil, independent of seed source presence or absence (Granström 1982, Staaf et al. 1987, Augusto et al. 2001, Jaroszewicz 2013). In samples taken at a depth of 5-10 cm in deciduous and coniferous forests 3 and 71 birch seeds per m² were detected (Jaroszewicz 2013, Jedrzejczak 2013). Regardless of the occurrence of seed trees, an average of 33 seeds per m² were present in the mineral soil down to a depth of 10-20 cm (Bossuyt et al. 2002), which may lead one to assume that birch seeds live longer in the soil than is often assumed (Fig. 5.2). With increasing soil depth, the number of viable birch seeds declined, but remained high enough for reforestation. On succession sites, where many seed sources are available for seed supply, 324 seeds per m² were recorded in the humus and litter layer (Hill & Stevens 1981). At depths of 0-5 cm and 5-10 cm in the mineral soils of these sites, densities reached 79-2880 seeds per m², and 20-880 seeds per m², respectively (Hill & Stevens 1981, Bakker et al. 1996a, Kalamees & Zobel 1998).

Irrespective of the timing during the year of soil sample collection, high densities of viable birch seeds could always be found. Highest seed densities of 20-3850 viable birch seeds per m² occur in the period from May to June (Warr et al. 1994). The results of the studies showed that seed densities were more dependent on the presence or absence of seed sources than on the timing of soil sampling or on the forest community of the sampling site (Kjellsson 1992, Houle 1998, Bossuyt & Hermy 2001).

The different studies, and the different assessments of birch seed longevity, explained the varying assignments of birch to the contrasting soil seed bank types, which ranged from purely transient (Bekker et al. 2000), through transient/short-term persistent (Thompson et al. 1997) to short-term/long-term persistent (Bakker et al. 1996a). The assumption made by Olmsted & Curtis (1947), Bakker et al. (1996b) and Graber & Thompson (1978) that seed rain from outside of a stand is necessary for the regeneration of the species where birch seed trees are not present on a site, due to an insufficient seed bank, cannot be supported without new research.

5.5.2 Soil seed bank of *Alnus glutinosa* (L.) GAERTN.

Viable alder seeds were detected in the soil seed bank less often than birch seeds (10 % of sample plots). This is probably due to the lower frequency of alder trees than birch in European managed forests. Seeds were only found in soils where there were seed sources close by (Kjellsson 1992, Falińska 1999, Decocq et al. 2004 - Fig. 5.2). Alder seeds were mainly detected on meadow and hayfield succession sites aged between 0-25 years. The seed density ranged between 8 and 216 seeds per m². The maximum was found in a 20-year-old dry hayfield. In contrast to this, the maximum number recorded on a wet hayfield succession site was only 80 seeds per m² (Bekker et al. 2000). In 40- to 175-year-old deciduous forests, 2-7 alder seeds per m² were observed (Kjellsson 1992, Decocq et al. 2004). The highest alder seed density recorded in humus and mineral soil was 354 seeds per m², which was obtained from a mixed lime-alder-birch forest (Kjellsson 1992). No alder seeds were detected in coniferous or mixed stands. The studies presenting the findings from such stands provided no information about the presence of alder seed trees or woodlands, in contrast to soil seed bank studies undertaken in deciduous stands or on succession sites. Apart from the study by Decocq et al. (2004), no viable alder seeds were found in soil samples taken between May and December; not even from samples taken next to alder seed trees (Kalamees & Zobel 1998, Warr et al. 1994). This is a clear indication of a transient seed bank.

The number of alder seeds transported vertically in the soil, and the depth of transport, could not be derived in any detail from the studies evaluated. Kalamees & Zobel (1998), who collected samples without litter from a pioneer forest with alder and birch, detected high numbers of viable birch seeds but no viable alder seeds. Only in one case very low densities of 2.4 and 3.2 viable alder seeds per m² were confirmed in two mineral soil samples taken close to seed trees in June (Decocq et al. 2004). However, studies providing the occurrence of alder seeds in the soil indicated that they tend to be more prevalent in the upper soil and in the humus layer than in the lower soil layers (Fig. 5.2). Kjellsson (1992) concluded, therefore, that alder seeds are short-lived and that large seed numbers in the soil were probably due to recent seed rain.

The buried seed experiment by Granström (1987) indicated a shorter lifespan of alder seeds than for birch. Early germination in the field before sampling could not be ruled out, but after 1.5 years of seed storage in soil the proportion of viable seeds was about 60 %, and only 2 %

after 5 years. Interestingly, the pericarp and wings of buried alder seeds were still intact, while parts of birch seeds had begun to decompose (Granström 1987). It seems unlikely that after a long period of vertical drift many alder seeds reach the deeper soil layers in a condition allowing for germination. Decocq et al. (2004) claimed that alder can establish a more persistent seed bank, whereas Thompson et al. (1997), Bekker et al. (2000) and Onaindia & Amezaga (2000) assigned alder to the transient seed bank type. Considering the lack of available data and literature, a reliable statement on the alder soil seed bank type is not possible. Some results suggested a transient seed bank, but this seems to have been influenced in part by peculiarities of the sites in question. In future research, typical alder sites such as floodplains should be included in sampling.

5.5.3 Soil seed banks of *Salix* spp. and *Populus tremula* L.

The results provided on *Salix* spp. were often no more specific than a mere reference to the genus "willow". Therefore, it is not possible to discuss different willow species in detail. Despite the common assumption that willows have short-lived seeds (Junttila 1976, Niiyama 1990, Barsoum 2002, Young & Clements 2003), the genus was the second most abundant pioneer tree species in the papers analyzed, occurring in 17 % of all sample plots. Poplar seeds, morphologically similar to willow seeds, were almost always absent in soil seed banks (1 % of sample plots). Three European aspen seeds per m² were observed only once by Heinrichs (2010) on a succession site, which indicates a rapid loss of poplar seed viability in soil (Worrell 1995, Barsoum 2002).

Viable willow seeds germinated predominantly in soil samples from succession sites, where seed trees were present. The highest recorded number of willow seeds was 350 seeds per m² on a 15-year-old meadow succession site dominated by willow (Falińska 1999). Summarizing all succession studies, seed density ranged from 6 to 350 seeds per m² (Bakker et al. 1996a, Falińska 1999, Bekker et al. 2000, Dölle & Schmidt 2009). A few willow seeds were also present in some soil samples from beech forests, with 7 and 28 seeds per m² (Staaf et al. 1987), in Norway spruce forests with 11 and 104 seeds per m² (Granström 1988, Berger et al. 2004), and in 65-year old mixed beech-spruce forest with 156 *S. caprea* seeds per m² (Berger et al. 2004). All of these authors studied the humus and mineral soil layers. In one study, *S. alba* L. grew at high frequencies in the vegetation (7-24 %), but no viable seeds were identified in the soil (Bissels et al. 2005). Gurnell et al. (2006) also detected willow species in the vegetation along a newly created riverbank but not in the seed bank. The authors explained the results by assuming transience of the seeds and immediate germination after ground contact.

Information on willow seed densities in different soil layers and at different depths was rare in the evaluated studies. Staaf et al. (1987) observed 7-14 willow seeds per m² in the humus layer and in the mineral soil at a depth of 0-5 cm in a beech forest. Bakker et al. (1996a), by contrast, recorded 80 goat willow seeds per m² in some 0-5 cm soil samples taken from 20- to 80-year old *Juniperus* shrubland, whereas 80 seeds per m² were detected in a 5-10 cm soil sample in annually grazed *Juniperus* shrubland. Six seeds of *S. caprea* were present only once in mineral soil sampled from a succession site (Dölle & Schmidt 2009). All authors reported the absence of willow seed trees, which highlights long willow seed dispersal distances (Schirmer 2006) and the possibility of formation of a willow soil seed bank (Fig. 5.2). However, viable willow and poplar seeds were only derived from samples collected in March and April (Staaf et al. 1987, Granström 1988, Bakker et al. 1996a, Falińska 1999, Bekker et al. 2000, Berger et al. 2004, Dölle & Schmidt 2009). Samples taken near willow and European aspen seed trees in May and June contained no viable seeds of either genus (Falińska 1999, Decocq et al. 2004) due to a rapid loss of germination ability after deposition on soil (Barsoum 2002). This also provides a strong indication of a transient seed bank for both genera.

Information about willow seeds artificially buried in the soil could not be found in the literature. Thompson et al. (1998) concluded that willows do not have a persistent seed bank. Certain aspen and *Salix* spp., especially *S. caprea*, an early flowering species, were assigned to the transient seed bank type (Thompson et al. 1997, 1998, Bekker et al. 2000), whereas *S. alba*, a late flowering willow species, was assigned to the long-term persistent seed bank category (Berger et al. 2004). Perhaps the high numbers of viable willow seeds recorded in the soil had all fallen into cracks or were from late-flowering willow species. It also seems possible that willow seeds protected by soil are viable for a longer time than those on the ground surface or humus layer, or that the movement of the smaller and lighter seeds to deeper soil layers proceeds rapidly, as documented by van Tooren (1988) and Burmeier et al. (2010).

5.5.4 Soil seed bank of *Sorbus aucuparia* L.

In contrast to all the other pioneer tree species mentioned, rowan was not found in any of the soil seed bank studies, although seed trees were present in some of the study areas (Granström 1982, Decocq et al. 2004, Dölle & Schmidt 2009, Heinrichs 2010, Jędrzejczak 2013). Often the only indication of successful reproduction was the presence of young rowan trees in the herb and shrub layer, for example, in conifer forests (Granström 1982, Heinrichs 2010). With secondary dormancy, the seeds can be part of the soil seed bank for at least 1-2 years (Leder 1992). However, the findings of this review indicate that rowan seeds are always absent from the soil seed bank. Grime et al. (1988 cited in Raspé et al. 2000) and Dölle & Schmidt (2009) assigned rowan to the transient seed bank type, because seeds persist in soil for less than 1 year. In contrast, Hill (1979), Leder (1992) and Erlbeck (1998) agreed that rowan seeds can remain viable in the soil for long time, up to 5 years. Based on the above, it would appear possible that rowan has a short-term persistent seed bank. This assumption would seem to have been confirmed by an experiment with buried pomes, which showed that more than 80-90 % of the seeds remain viable after 2 years storage in the soil. During the third year the ability to germinate decreased to 30-50 %, but some rowan seeds remained viable after 5 years of storage (Granström 1987). Up to 9 % of fresh, stratified rowan seeds exposed under field conditions germinated in the second or third year after sowing (Miles 1974). Due to the clumped distribution of rowan seeds by birds (McDonnell & Stiles 1983), future studies of the occurrence of rowan in soil seed banks should take into consideration the structural elements used as perches by birds (Tiebel et al. 2017).

5.6 Conclusions

Pioneer tree species are short-lived, light demanding species, which are very important for the successful colonization and reforestation of large disturbed woodlands in central and northwest Europe. Soil seed banks can drive reforestation in the absence of seed rain. Soil seed banks in woodlands play an important role in succession and in the regeneration of disturbed areas in European temperate forests.

This review showed that pioneer tree species do not possess the kind of long-term seed banks that certain herbaceous species can have. The findings suggest that birch is the only pioneer tree species of temperate forests in central and northwest Europe with a longer-lived soil seed bank. Often birch was the second most abundant species in soil and the only pioneer tree species with a high degree of consistency in soil seed banks. In medium to deeper soil layers (5-10 cm) birch seeds seem to have at least a short-term persistent seed bank. Alder seeds are poorly represented in forest soils compared to birch, so a reliable statement on alder soil seed bank type is not possible; the few results available suggest a transient seed bank. The studies for willow and poplar seeds partly confirmed the assumption of very short-lived seeds, although willow was the second most abundant pioneer tree species in soil seed banks and also found in mineral soil (0-5 cm). Surprisingly in the case of rowan, the only fleshy-fruited pioneer tree species with proven seed dormancy, a transient seed bank must be assumed due to the absence of rowan seeds in the soil. Buried seed experiments showed, however, that rowan seeds can build up a short-term persistent seed bank due to dormancy.

Statements on the seed densities of pioneer tree species in the soils of different coniferous and deciduous forest types cannot be given. The reason for this is that these seed densities are primarily influenced by the number of and the distance from seed sources, and the seasons of seed production and seed dispersal. Our review revealed that the successful regeneration of birch, alder and willow depends mainly on the proximity of seed trees. Therefore, the proximity of trees is important for the regeneration of species with short-lived seeds. The findings of the review also indicate a dependence between seed density in the soil and the season in which soil sampling occurs. Maximum seed densities of birch, alder and willow were detected during and shortly after seed rain. No statement can be made in this regard in relation to rowan and poplar.

This review revealed a number of open questions concerning the capacity of all European pioneer tree species to establish seed banks. These issues are connected to: (a) the seed viability under different soil conditions and litter thickness; (b) the speed of seed movement into deeper soil layers; and (c) the direct correlation between the proximity of seed trees and the resultant number of seeds in the soil. At present, it can be concluded that birch, representative of pioneer tree species in temperate forests of central and northwest Europe, has the capacity to establish a seed bank of a duration of 1-5 years, sufficient to compensate for years with lower levels of seed production and to regenerate successfully after disturbance. However, the soil seed bank must be supplemented by fresh seeds from surrounding seed trees as often as possible in order to guarantee continuous regeneration.

5.7 References

Amezaga I, Onaindia M (1997). The effect of evergreen and deciduous coniferous plantations

on the field layer and seed bank of native woodlands. Ecography 20: 308-318. - doi: 10.1111/j.1600-0587.1997.tb00375.x

Amrein D, Rusterholz H-P, Baur B (2005). Disturbance of suburban *Fagus* forests by recreational activities: effects on soil characteristics, aboveground vegetation and seed bank. Applied Vegetation Science 8: 175-182. - doi: 10.1658/1402-2001(2005)008[0175:DOSFFB]2.0.CO;2

Argus GW (2006). Guide to *Salix* (willow) in the Canadian Maritime Provinces (New Brunswick, Nova Scotia, and Prince Edward Island). Canadian Museum of Nature, Ottawa, Canada, pp. 49. [online] URL: http://accs.uaa.alaska.edu/files/botany/publications/2006/GuideSalixCanadianAtlanticMaritime.pdf

Atkinson MD (1992). *Betula pendula* Roth (*B. verrucosa* Ehrh.) and *B. pubescens* Ehrh. Journal of Ecology 80: 837-870. - doi: 10.2307/2260870

Augusto L, Dupouey J-L, Picard J-F, Ranger J (2001). Potential contribution of the seed bank in coniferous plantations to the restoration of native deciduous forest vegetation. Acta Oecologica 22: 87-98. - doi: 10.1016/S1146-609X(01)01104-3

Bakker JP, Bakker ES, Rosén E, Verwej GL, Bekker RM (1996a). Soil seed bank composition along a gradient from dry alvar grassland to *Juniperus* shrubland. Journal of Vegetation Science 7: 165-176. - doi: 10.2307/3236316

Bakker JP, Poschlod P, Strykstra RJ, Bekker RM, Thompson K (1996b). Seed banks and seed dispersal: important topics in restoration ecology. Acta Botanica Neerlandica 45: 461-490. - doi: 10.1111/j.1438-8677.1996.tb00806.x

Barnes BV, Zak DR, Denton SR, Spurr SH (1998). Forest ecology. John Wiley and Sons, New York, USA, pp. 774. Barsoum N (2002). Relative contributions of sexual and asexual regeneration strategies in *Populus nigra* and *Salix alba* during the first years of establishment on a braided gravel bed river. Evolutionary Ecology 15: 255-279. - doi: 10.1023/A:1016028730129

Bekker RM, Verwej GL, Bakker JP, Fresco LFM (2000). Soil seed bank dynamics in hayfield succession. Journal of Ecology 88: 594-607. - doi: 10.1046/j.1365-2745.2000.00485.x

Berger TW, Sun B, Glatzel G (2004). Soil seed banks of pure spruce (*Picea abies*) and adjacent mixed species stands. Plant and Soil 264: 53-67. - doi: 10.1023/B:PLSO.0000047753.36424.41

Bissels S, Donath TW, Hölzel N, Otte A (2005). Ephemeral wetland vegetation in irregularly flooded arable fields along the northern Upper Rhine: the importance of persistent seedbanks. Phytocoenologia 35: 469-488. - doi: 10.1127/0340-269X/2005/0035-0469

Bjorkbom JC (1967). Seedbed-preparation methods for paper birch. Research Paper NE-209, Northeastern Forest Experiment Station, USDA Forest Service, Upper Darby, USA, pp. 15.

Bjorkbom JC (1971). Production and germination of paper birch seed and its dispersal into a forest opening. Research Paper NE-209, Northeastern Forest Experiment Station, USDA Forest Service, Upper Darby, USA, pp. 15.

Bossuyt B, Hermy M (2001). Influence of land use history on seed banks in European temperate forest ecosystems: a review. Ecography 24: 225-238. - doi: 10.1034/j.1600-0587.2001.240213.x

Bossuyt B, Heyn M, Hermy M (2002). Seed bank and vegetation composition of forest stands of varying age in central Belgium: consequences for regeneration of ancient forest vegetation. Plant Ecology 162: 33-48. - doi: 10.1023/A:1020391430072

Bossuyt B, Honnay O (2008). Can the seed bank be used for ecological restoration? An overview of seed bank characteristics in European communities. Journal of Vegetation Science 19: 875-884. - doi: 10.3170/2008-8-18462

Bradbeer JW (1988). Seed dormancy and germination. Blackie and Son Ltd., Glasgow and London, UK, pp. 146.

Brouwer W, Stählin A (1975). Handbuch der Samenkunde für Landwirtschaft, Gartenbau und Forstwirtschaft [Seed manual of agriculture, horticulture and forestry]. DLG-Verlags-GmbH, Frankfurt am Main, Germany, pp. 655. [in German]

Brown AHF, Oosterhuis L (1981). The role of buried seed in coppicewood. Biological Conservation 21: 19-38. - doi: 10.1016/0006-3207(81)90066-5

Buckley GP, Howell R, Anderson MA (1997). Vegetation succession following ride edge management in lowland plantations and woods. 2. The seed bank resource. Biological Conservation 82: 305-316. - doi: 10.1016/S0006-3207(97)00025-6

Burmeier S, Eckstein RL, Otte A, Donath TW (2010). Desiccation cracks act as natural seed traps in flood-meadow systems. Plant and Soil 333: 351-364. - doi: 10.1007/s11104-010-0350-1

Chmelar J, Meusel W (1986). Die Weiden Europas [Willows of Europe]. A. Ziemsen Verlag, Wittenberg, Germany, pp. 144. [in German]

Clark JS, Macklin E, Wood L (1998). Stages and spatial scales of recruitment limitation in southern Appalachian forests. Ecological Monographs 68: 213-235. - doi: 10.1890/0012-9615(1998)068[0213:SASSOR]2.0.CO;2

Decocq G, Valentin B, Toussaint B, Hendoux F, Saguez R, Bardat J (2004). Soil seed bank composition and diversity in a managed temperate deciduous forest. Biodiversity and Conservation 13: 2485-2509. - doi: 10.1023/B:BIOC.0000048454.08438.c6

Dölle M, Schmidt W (2009). The relationship between soil seed bank, above-ground vegetation and disturbance intensity on old-field successional permanent plots. Applied Vegetation Science 12: 415-428. - doi: 10.1111/j.1654-109X.2009.01036.x

Donelan M, Thompson K (1980). Distribution of buried viable seeds along a successional series. Biological Conservation 17: 297-311. - doi: 10.1016/0006-3207(80)90029-4

Dougall TAG, Dodd JC (1997). A study of species richness and diversity in seed banks and its use for the environmental mitigation of a proposed holiday village development in a coniferized woodland in south east England. Biodiversity and Conservation 6: 1413-1428. - doi: 10.1023/A:1018345915418

Ebrecht L, Schmidt W (2008). Bedeutung der Bodensamenbank und des Diasporentransports durch Forstmaschinen für die Entwicklung der Vegetation auf Rückegassen [Impact of soil seed bank and diaspore transportation by forest machines on the development of vegetation along skid trails]. Forstarchiv 79: 91-105. [in German]

Egawa C, Tsuyuzaki S (2013). The effects of litter accumulation through succession on seed bank formation for small and large seeded species. Journal of Vegetation Science 24: 1062-1073. - doi: 10.1111/jvs.12037

Erlbeck R (1998). Die Vogelbeere (*Sorbus aucuparia*) - ein Porträt des Baumes des Jahres 1997 [Rowan (*Sorbus aucuparia*) - a portrait of the tree of the year 1997]. In: "Beiträge zur Vogelbeere" [Article about rowan] (Schmidt O ed.). Berichte aus der Bayrischen Landesanstalt für Wald und Forstwirtschaft 17: pp. 2-14. [in German]

Falińska K (1999). Seed bank dynamics in abandoned meadows during a 20-year period in the Bialowieza National Park. Journal of Ecology 87: 461-475. - doi: 10.1046/j.1365-2745.1999.00364.x

Fenner M (1985). Seed ecology. Chapman and Hall Ltd, London and New York, UK and USA, pp. 151.

Fischer H, Huth F, Hagemann U, Wagner S (2016). Developing restoration strategies for temperate forests using natural regeneration processes. In: "Restoration of boreal and temperate forests" (Stanturf JA ed.). CRC Press, Boca Raton, USA, pp. 103-164.

Gage EA, Cooper DJ (2005). Patterns of willow seed dispersal, seed entrapment, and seedling establishment in a heavily browsed montane riparian ecosystem. Canadian Journal of Botany 83: 678-687. - doi: 10.1139/B05-042

Godefroid S, Phartyal SS, Koedam N (2006). Depth distribution and composition of seed banks under different tree layers in a managed temperate forest ecosystem. Acta Oecologica 29: 283-292. - doi: 10.1016/j.actao.2005.11.005

Graber RE, Thompson DF (1978). Seeds in the organic layers and soil of four beech-birchmaple stands. Research Paper NE-401, Northeastern Forest Experiment Station, USDA Forest Service, Broomall, USA, pp. 10. [online] URL: http://www.fs.fed.us/ne/newtown_square/publications/research_papers/pdfs/scanned/OCR/ne_rp401.pdf

Grandin U (2001). Short-term and long-term variation in seed bank/vegetation relations along an environmental and successional gradient. Ecography 24: 731-741. - doi: 10.1111/j.1600-0587.2001.tb00534.x

Granström A (1982). Seed banks in five boreal forest stands originating between 1810 and 1963. Canadian Journal of Botany 60: 1815-1821. - doi: 10.1139/b82-228

Granström A, Fries C (1985). Depletion of viable seeds of *Betula pubescens* and *Betula verrucosa* sown onto some north Swedish forest soils. Canadian Journal of Forest Research 15: 1176-1180. - doi: 10.1139/x85-191

Granström A (1987). Seed viability of fourteen species during five years of storage in a forest soil. Journal of Ecology 75: 321-331. - doi: 10.230 7/2260421

Granström A (1988). Seed banks at six open and afforested heathland sites in southern Sweden. Journal of Applied Ecology 25: 297-306. - doi: 10.2307/2403627

Greene DF, Johnson EA (1996). Wind dispersal of seeds from a forest into a clearing. Ecology 77: 595-609. - doi: 10.2307/2265633

Gurnell AM, Boitsidis AJ, Thompson K, Clifford NJ (2006). Seed bank, seed dispersal and vegetation cover: Colonization along a newly-created river channel. Journal of Vegetation Science 17: 665-674. - doi: 10.1111/j.1654-1103.2006.tb02490.x

Heinrichs S (2010). Response of the understorey vegetation to select cutting and clear cutting in the initial phase of Norway spruce conservation. PhD book, Faculty of Mathematics and Natural Sciences, Georg-August-University Göttingen, Göttingen, Germany, pp. 177. [online] URL: http://ediss.uni-goettingen.de/handle/11858/00-1735-0000-0006-B691-E

Hester AJ, Gimingham CH, Miles J (1991). Succession from heather moorland to birch woodland. III. Seed availability, germination and early growth. Journal of Ecology 79: 329-344. - doi: 10.2307/2260716

Hill MO (1979). The development of a flora in even-aged plantations. In: "The ecology of even-aged forest plantations" (Ford ED, Malcolm DC, Atterson J eds). Institute of Terrestrial Ecology, Cambridge, UK, pp. 175-192. [online] URL: http://nora.nerc.ac.uk/id/eprint/7075/1/N007075CP.pdf

Hill MO, Stevens PA (1981). The density of viable seed in soils of forest plantations in upland Britain. Journal of Ecology 69: 693-709. - doi: 10.2307/2259692

Holm SO (1994). Reproductive patterns of *Betula pendula* and *B. pubescens* coll. along a regional altitudinal gradient in northern Sweden. Ecography 17: 60-72. - doi: 10.1111/j.1600-0587.1994.tb00077.x

Hopfensperger KN (2007). A review of similarity between seed bank and standing vegetation across ecosystems. Oikos 116: 1438-1448. - doi: 10.1111/j.2007.0030-1299.15818.x

Houle G (1998). Seed dispersal and seedling recruitment of *Betula alleghaniensis*: spatial inconsistency in time. Ecology 79: 807-818. - doi: 10.1890/0012-9658(1998)079[0807:SDASRO]2.0.CO;2

Huth F (2009). Untersuchungen zur Verjüngungsökologie der Sand-Birke (*Betula pendula* Roth) [Study of regeneration cycle of sand birch (*Betula pendula* Roth)]. PhD book, Faculty of Forest, Geo and Hydro, Dresden University of Technology, Dresden, Germany, pp. 383. [in German]

Hynynen J, Niemistö P, Viherä-Aarnio A, Brunner A, Hein S, Velling P (2010). Silviculture of birch (*Betula pendula* Roth and *Betula pubescens* Ehrh.) in northern Europe. Forestry 83: 103-119. - doi: 10.1093/forestry/cpp035

Jankowska-Blaszczuk M (1998). Variability of the soil seed banks in the natural deciduous forest in the Bialowieza National Park. Acta Societatis Botanicorum Poloniae 67: 313-324. - doi: 10.5586/asbp.1998.040

Jankowska-Blaszczuk M, Kwiatkowska AJ, Panufnik D, Tanner E (1998). The size and diversity of the soil seed banks and the light requirements of the species in sunny and shady natural communities of the Bialowieza Primeval Forest. Plant Ecology 136: 105-118. - doi: 10.1023/A:1009750201803

Jaroszewicz B (2013). Endozoochory by European bison influences the build-up of the soil seed bank in subcontinental coniferous forest. European Journal of Forest Research 132: 445-452. - doi: 10.1007/s10342-013-0683-4

Jedrzejczak E (2013). Soil seed bank in selected patches of vegetation in the beech forest of Beskid Åšlaski (southern Poland). Casopis Slezského Zemského Muzea 62: 137-150. - doi: 10.2478/cszma-2013-0015

Junttila O (1976). Seed germination and viability in five *Salix* species. Astarte 9: 19-24.

Kalamees R, Zobel M (1998). Soil seed bank composition in different successional stages of a species rich wooded meadow in Laelatu, western Estonia. Acta Oecologica 19: 175-180. - doi: 10.1016/S1146-609X(98)80021-0

Karlsson M (2001). Natural regeneration of broadleaved tree species in southern Sweden - Effects of silvicultural treatments and seed dispersal from surrounding stands. PhD book, Southern Swedish Forest Research Centre, Swedish University of Agricultural Sciences, Alnarp, Sweden, pp. 44. [online] URL: http://pub.epsilon.slu.se/42/

Karrenberg S, Kollmann J, Edwards PJ (2002). Pollen vectors and inflorescence morphology in four species of *Salix*. Plant Systematics and Evolution 235: 181-188. - doi: 10.1007/s00606-002-0231-z

Karrenberg S, Suter M (2003). Phenotypic tradeoffs in the sexual reproduction of Salicaceae from flood plain. American Journal of Botany 90: 749-754. - doi: 10.3732/ajb.90.5.749

Kjellsson G (1992). Seed banks in Danish deciduous forests: species composition, seed influx and distribution pattern in soil. Ecography 15: 86-100. - doi: 10.1111/j.1600-0587.1992.tb00012.x

Komulainen M, Vieno M, Yarmishko VT, Daletskaja TD, Maznaja EA (1994). Seedling establishment from seeds and seed banks in forests under long-term pollution stress: a potential for vegetation recovery. Canadian Journal of Botany 72: 143-149. - doi: 10.1139/b94-019

Lautenschlager D (1994). Die Weiden von Mittelund Nordeuropa [The willowe of central and northern Europe]. Birkhäuser Verlag, Basel, Switzerland, pp. 171. [in German]

Leck MA, Parker VT, Simpson RL (1989). Ecology of soil seed banks. Academic Press Limited, London, UK, pp. 462.

Leck MA, Parker VT, Simpson RL (2008). Seedling ecology and evolution. University Press, Cambridge, UK, pp. 514. [online] URL: http://books.google.com/books?id =lPSYJwMHanAC

Leder B (1992). Weichlaubhölzer: Verjüngungsökologie, Jugendwachstum und Bedeutung in Jungbeständen der Hauptbaumarten Buche und Eiche [Pioneer tree species: regeneration, youth growth and importance in young stands of the main tree species beech and oak]. Landesanstalt für Forstwirtschaft Nordrhein-Westfalen, Arnsberg, Germany, pp. 416. [in German]

Ludwig JW, Bunting ES, Harper JL (1957). The influence of environment on seed and seedling mortality: III. The influence of aspect on maize germination. Journal of Ecology 45: 205-224. - doi: 10.2307/2257085

McDonnell MJ, Stiles EW (1983). The structural complexity of old field vegetation and the recruitment of bird-dispersed plant species. Oecologia 56: 109-116. - doi: 10.1007/BF00378225

McVean DN (1953). *Alnus glutinosa* (L.) Gaertn. Journal of Ecology 41: 447-466. - doi: 10.2307/2257070

McVean DN (1956). Ecology of *Alnus glutinosa* (L.) Gaertn.: VI. Post-glacial history. Journal of Ecology 44: 331-333. - doi: 10.2307/2256825

Milberg P (1995). Soil seed bank after eighteen years of succession from grassland to forest. Oikos 72: 3-13. - doi: 10.2307/3546031

Miles J (1974). Experimental establishment of new species from seed in Callunetum in north-east Scotland. Journal of Ecology 62: 527-551. - doi: 10.2307/2258997

Miller GR, Cummins RP (2003). Soil seed banks of woodland, heathland, grassland, mire and montane communities, Cairngorm Mountains, Scotland. Plant Ecology 168: 255-266. - doi: 10.1023/A:1024464028195

Mitschell RJ, Marrs RH, Auld MHD (1998). A comparative study of the seedbanks of heathland and successional habitats in Dorset, southern England. Journal of Ecology 86: 588-596. - doi: 10.1046/j.1365-2745.1998.00281.x

Niiyama K (1990). The role of seed dispersal and seedling traits in colonization and coexistence of *Salix* species in a seasonally flooded habitat. Ecological Research 5: 317-331. - doi: 10.1007/BF02347007

Olmsted NW, Curtis JD (1947). Seeds of the forest floor. Ecology 28: 49-52. - doi: 10.2307/1932917

Onaindia M, Amezaga I (2000). Seasonal variation in the seed banks of native woodland and coniferous plantations in northern Spain. Forest Ecology and Management 126: 163-172. - doi: 10.1016/S0378-1127(99)00099-7

Osumi K, Sakurai S (1997). Seedling emergence of *Betula maximowicziana* following human disturbance and the role of buried viable seeds. Forest Ecology and Management 93: 235-243. - doi: 10.1016/S0378-1127(96)03963-1

Perala DA, Alm AA (1990). Reproductive ecology of birch: a review. Forest Ecology and Management 32: 1-38. - doi: 10.1016/0378-1127(90)90104-J

Pietzarka U, Roloff A (2010). *Alnus glutinosa* (L.) Gaertn. In: "Bäume Mitteleuropas" [Trees of central Europe] (Roloff A, Weisberger H, Lang U, Stimm B eds). Wiley-VCH Verlag GmbH and Co. KGaA, Weinheim, Germany, pp. 141-155. [in German]

Plue J, Verheyen K, Calster H, Marage D, Thompson K, Kalamees R, Jankowska-Blaszczuk M, Bossuyt B, Hermy M (2010). Seed banks of temperate deciduous forests during secondary succession. Journal of Vegetation Science 21: 965-978. - doi: 10.1111/j.1654-1103.2010.01203.x

Raspé O, Findlay C, Jacquemart A-L (2000). *Sorbus aucuparia* L. Journal of Ecology 88: 910-930. - doi: 10.1046/j.1365-2745.2000.00502.x

Rydgren K, Hestmark G, Okland RH (1998). Revegetation following experimental disturbance in a boreal old-growth *Picea abies* forest. Journal of Vegetation Science 9: 763-776. - doi: 10.2307/3237042

Saatkamp A, Poschlod P, Venable DL (2014). The functional role of soil seed banks in natural communities. In: "Seeds: The Ecology of Regeneration in Plant Communities" (Gallagher RS ed). CAB International, Oxfordshire, UK, pp. 263-295. [online] URL: http://books.google.com/books?id=3Bb4AgAAQBAJ

Sarvas R (1952). On the flowering of birch and the quality of seed crop. Communicationes Instituti Forestalis Fenniae 40: 1-35.

Schirmer R (2006). *Salix alba* Linné. In: "Enzyklopädie der Laubbäume" [Encyclopedia of deciduous trees] (Schütt P, Weisberger H, Schuck HJ, Lang U, Stimm B eds). Nikol Verlagsgesellschaft mbH & Co KG, Hamburg, Germany, pp. 535-550. [in German]

Schölch M (1998). Zur natürlichen Widerbewaldung ohne forstliche Steuerung [Natural reforestation without forestry interventions]. Schriftenreihe Freiburger Forstlicher Forschung 1, Forstliche Versuchs- und Forschungsanstalt Baden-Württemberg, Freiburg, Germany, pp. 245. [in German]

Seidl R, Schelhaas M-J, Rammer W, Verkerk PJ (2014). Increasing forest disturbances in Europe and their impact on carbon storage. Nature Climate Change 4: 806-810. - doi: 10.1038/nclimate2318

Simpson RL, Leck MA, Parker VT (1989). Seed banks: general concepts and methodological issues. In: "Ecology of Soil Seed Banks" (Leck MA, Parker VT, Simpson RL eds). Academic Press Limited, London, UK, pp. 3-8.

Skoglund J, Verwijst T (1989). Age structure of woody species populations in relation to seed rain, germination and establishment along the river Dalälven, Sweden. Vegetation 82: 25-34. - doi: 10.1007/BF00217979

Sperens U (1997). Long-term variation in, and effects of fertiliser addition on, flower, fruit and seed production in the tree *Sorbus aucuparia* (Rosaceae). Ecography 20: 521-534. - doi: 10.1111/j.1600-0587.1997.tb00421.x

Spethmann W (2000). Generative Gehölzvermehrung [Generative propagation of woody plants]. In: "Krüssmanns Gehölzvermehrung" [Kruessmann's propagation of woody plants] (Mac Cárthaigh D, Spethmann W eds). Parey Verlag, Berlin, Germany, pp. 2-57. [in German]

Staaf H, Jonsson M, Olsén L-G (1987). Buried germinative seeds in mature beech forests with different herbaceous vegetation and soil types. Holarctic Ecology 10: 268-277. - doi: 10.1111/j.1600-0587.1987.tb00768.x

Sullivan KA, Ellison AM (2006). The seed bank of hemlock forests: implications for forest regeneration following hemlock decline. The Journal of the Torrey Botanical Society 133: 393-402. - doi: 10.3159/1095-5674(2006)133[393:TSBOHF]2.0.CO;2

Thompson K, Grime JP (1979). Seasonal variation in the seed banks of herbaceous species in ten contrasting habitats. Journal of Ecology 67: 893-921. - doi: 10.2307/2259220

Thompson K, Bakker J, Bekker R (1997). The soil seed banks of North West Europe: methodology, density and longevity. Cambridge University Press, Cambridge, UK, pp. 276. [online] URL: http://books.google.com/books?id=zGDcH8f5n-wC

Thompson K, Bakker JP, Bekker RM, Hodgson JG (1998). Ecological correlates of seed persistence in soil in the north-west European flora. Journal of Ecology 86: 163-169. - doi: 10.1046/j.1365-2745.1998.00240.x

Tiebel K, Karge A, Huth F, Wehnert A, Wagner S (2017). Strukturelemente fördern die Samenausbreitung durch Vögel [Structural elements promote seed dispersal by birds]. AFZ-Der Wald 20: 24-27. [in German]

Van Tooren BF (1988). The fate of seeds after dispersal in chalk grassland: the role of the bryophyte layer. Oikos 53: 41-48. - doi: 10.2307/3565661

Warr SJ, Kent M, Thompson K (1994). Seed bank composition and variability in five woodlands in south-west England. Journal of Biogeography 21: 151-168. - doi: 10.2307/2845469

Worrell R (1995). European aspen (*Populus tremula* L.): a review with particular reference to Scotland: I. Distribution, ecology and genetic variation. Forestry 68: 93-105. - doi: 10.1093/forestry/68.2.93

Young JA, Clements CD (2003). Seed germination of willow species from a desert riparian ecosystem. Journal of Range Management 56: 496-500. - doi: 10.2307/4003842

Zerbe S (2001). On the ecology of *Sorbus aucuparia* (Rosaceae) with special regard to germination, establishment and growth. Polish Botanical Journal 46: 229-239.

Zerbe S (2009). Renaturierung von Waldökosystemen [Restoration of forest ecosystems]. In: "Renaturierung von Ökosystemen in Mitteleuropa" [Restoration of ecosystems in central Europe] (Zerbe S, Wiegleb G eds). Spektrum Akademischer Verlag, Heidelberg, Germany, pp. 153-182. [in German]

Zobel M, Kalamees R, Püssa K, Roosaluste E, Moora M (2007). Soil seed bank and vegetation in mixed coniferous forest stands with different disturbance regimes. Forest Ecology and Management 250: 71-76. - doi: 10.1016/j.foreco.2007.03.011

Żywiec M, Ledwoń M (2008). Spatial and temporal patterns of rowan (*Sorbus aucuparia* L.) regeneration in West Carpathian subalpine spruce forest. Plant Ecology 194: 283-291. - doi: 10.1007/s11258-007-9291-z

Żywiec M, Holeksa J, Wesolowska M, Szewczyk J, Zwijacz-Kozica T, Kapusta P (2013). *Sorbus aucuparia* regeneration in a coarse-grained spruce forest - a landscape scale. Journal of Vegetation Science 24: 735-743. - doi: 10.1111/j.1654-1103.2012.01493.x

Supplementary material

Table S1 - Summary of the 33 seed bank studies in central and north-west European temperate forests selected.

Author(s)	Date	Country	Forest type	Age [yr]
Amrein et al.	2005	Switzerland	Deciduous forest	ancient
Augusto et al.	2001	France, northeast	Beech forest	57
			Oak forest	76 & 78
			Douglas fir plantation	65
			Pine plantation	43 & 65
			Spruce plantation	35 & 65
Bakker et al.	1996a	Sweden	Juniper shrubland on former grazing land	20
			Juniper shrubland on former grazing land	55
			Juniper shrubland on former grazing land	80
Bekker et al.	2000	Netherlands	Succession on dry and wet hayfield	7
			Succession on dry and wet hayfield	15
			Succession on dry and wet hayfield	20
			Succession on dry and wet hayfield	25
Berger et al.	2004	Austria	Secondary Norway spruce stand	53 - 65
			Mixed beech-spruce stand	89
Bossuyt et al.	2002	Belgium	Mixed beech-oak stand	55
			Mixed beech-oak stand	97
			Mixed beech-oak stand	116
Brown & Oosterhuis	1981	England, east	Abandoned coppice wood	30 - 40
Buckley et al.	1997	England, south	Beech plantation	36
			Beech plantation	52 - 58
			Oak plantation	40 - 55
			Corsican pine plantation	27 - 28
			Hazel coppice with oak standards	n.a.
			Hazel coppice with oak and ash standards	n.a.
			Ash-beech stand	mature
Decocq et al.	2004	France, north	Deciduous forest of former coppice with oak	ancient
Dölle & Schmidt	2009	Germany	Succession on former arable field	22
			Succession on former arable field	36
Donelan & Thompson	1980	England	Succession on former grassland	50
			Succession on former grassland	80
			Succession on former grassland	100
			Oak forest	200
Dougall & Dodd	1997	England, south-east	Conifer plantation	0 - 10
			Conifer plantation	10 - 20
			Conifer plantation	21 - 40
			Conifer plantation	> 65
			Broadleaf plantation	ancient
			Semi-natural broadleaf edge habitats	ancient
Ebrecht & Schmidt	2008	Germany	Beech forest	ancient
			Mixed beech forest	ancient
			Norway spruce stand	ancient

Author(s)	Date	Country	Forest type	Age [yr]
Falińska	1999	Poland, Belarus	Succession from meadow	0
			Succession from meadow	5
			Succession from meadow	10
			Succession from meadow	15
			Succession from meadow	20
Grandin	2001	Sweden	Succession from shoreline to mixed spruce-deciduous forest	mature
Granström	1982	Sweden, north	Pine stand	16
			Pine stand	29
			Pine stand	50
			Spruce stand with scattered pine	120
			Spruce stand with scattered pine	169
Granström	1988	Sweden	Norway spruce forest on former heathland	30
			Norway spruce forest on former heathland	35
			Norway spruce forest on former heathland	64
			Norway spruce forest on former heathland	73
			Clearfelled Norway spruce forest	85
Heinrichs	2010	Germany	Norway spruce forest	ancient
			Clearfelled Norway spruce forest	4
Hester et al.	1991	Scotland	Succession on heathland	17
			Succession on heathland	28
			Succession on heathland	36 - 37
			Succession on heathland	48
			Succession on heathland	63
Hill & Stevens	1981	England	Clearfelled Douglas fir plantation	4
			Sitka spruce plantation on former heathland	5 - 10
			Sitka spruce plantation on former heathland	17 - 18
			Sitka spruce plantation on former heathland	29 - 30
			Sitka spruce plantation on former heathland	30 - 31
			Sitka spruce plantation on former heathland	36 - 45
			Japanese larch plantation on former heathland	37
			Scots pine plantation on former heathland	37
			Oak wood	n.a.
Jankowska-Błaszczuk	1998	Poland, Belarus	Primary mixed forest	ancient
			Secondary deciduous forest	ancient
Jankowska-Błaszczuk et al.	1998	Poland, Belarus	Hornbeam forest	ancient
			Oak forest	ancient
Jaroszewicz	2013	Poland, Belarus	Spruce-pine stand	ancient
Jędrzejczak	2013	Poland, south	Beech forest	n.a.
Kalamees & Zobel	1998	Estonia	Succession on grassland	20
			Succession on grassland	long term
Kjellsson	1992	Denmark	Maple stand	40
			Beech forest	62 & 70
			Deciduous stand	141 & 145
			Mixed beech-oak stand	175
			Oak-ash stand of former hazel coppice forest	180
Komulainen et al.	1994	Russia, north	Pine forest	young
Milberg	1995	Sweden, south	Succession on grassland	18
Miller & Cummins	2003	Scotland	Oak-birch woodland	n.a.
			Pine woodland	n.a.

Author(s)	Date	Country	Forest type	Age [yr]
Mitschell et al.	1998	England	Succession on heathland	23
			Succession on heathland	30
			Succession on heathland	43
			Succession on heathland	48 - 49
Staaf et al.	1987	Sweden, south	Beech forest	90
			Beech forest	95
			Beech forest	100
			Beech forest	110
			Beech forest	140
			Beech forest	150
Thompson & Grime	1979	England	Deciduous forest	ancient
Warr et al.	1994	England, south-west	Cut oak coppice forest	8
			Cut oak coppice forest	24 - 25
			Abandoned hazel coppice forest	74
			Abandoned oak coppice forest	78
			Abandoned oak coppice forest	94
			Felled cherry and ash stand	2
			Oak stand	45
			Birch woodland on abandoned fields	80 - 90
			Oak woodland (abandoned coppice)	140
			Mixed conifer stand	18
			Hybrid larch stand	20
			Douglas fire stand	20 & 24
			Norway spruce stand	24
			Sitka spruce stand	28 - 29
			Sitka spruce stand	42
			Japanese larch stand	42
			European larch stand	53

Chapter 6

Do birch and rowan establish soil seed banks sufficient to compensate for a lack of seed rain after forest disturbance?

Katharina Tiebel[1], Franka Huth[1], Julian Heinrich[1], Sven Wagner[1] (submitted)

Plant and Soil

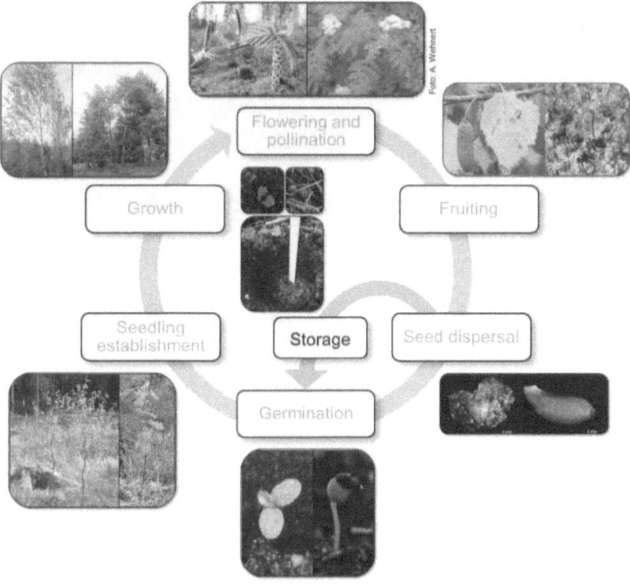

[1]TU Dresden, Institute of Silviculture and Forest Protection, Chair of Silviculture, Pienner Str. 8, 01737 Tharandt, Germany

6.2 Introduction

Soil seed banks are reservoirs of viable seeds stored in the soil for a period of time (Bossuyt and Honnay 2008; Leck et al. 2008). Changing environmental conditions can cause disturb-ances. These disturbances may in turn be followed by favorable conditions enabling seeds stored in the soil seed bank to germinate successfully (Bossuyt and Hermy 2001; Thompson et al. 1997). Soil seed banks also serve to preserve population genetics over time (Bossuyt and Honnay 2008; Hopfensberger 2007), but the duration of seed viability is species specific (Thompson et al. 1997). Whereas the seeds of some herbaceous species can remain viable for decades (Kjellsson 1992; Telewski and Zeevaart 2002), the seeds produced by most tree spe-cies are viable for only a few years. The trees with the longest known seed longevity in the soil are *Prunus serotina* Ehrh. and *Prunus pensylvanica* L.f., the seeds of which remain viable in the soil for 2-5 years and up to 30 years, respectively (Marquis 1975). Given the comparatively short viability of tree seeds, most previous soil seed bank studies have focused on herbaceous plants (e.g. Bernhardt and Poschlod 1992; Bossuyt and Honnay 2008; Hopfensperger 2007). Herbaceous plants are usually the largest group of plant species presented in the soil seed banks found in all soil types and ecosystems (see Jankowska-Błaszczuk 1998; Landenberger and McGraw 2004; Staaf et al. 1987; Zobel et al. 2007). Past studies of soil seed banks have also dealt with succession processes and landscape changes (see Granström 1988; Hopfensperger 2007; Kjellsson1992; Olmsted and Curtis 1947). Although Bossuyt and Honnay (2008) mentioned that soil seed banks are important sources of seeds and drivers of the regeneration of disturbed forest sites, tree species have received only minor consideration in previous soil seed bank studies, especially in studies focusing on central Europe.

The natural regeneration of disturbed sites in the temperate forests of Europe depends mostly on widespread early successional tree species, such as genera of *Betula* spp., *Salix* spp., *Populus* spp., *Alnus* spp. and *Sorbus aucuparia* L. (Argus 2006; Atkinson 1992; Zerbe 2001). The regeneration of pioneer tree species primarily follows a recent seed rain rather than seedling emergence from a soil seed bank (Buckley et al. 1997; Graber and Thompson 1978; Heinrichs 2010; Hill and Stevens 1981; Olmsted and Curtis 1947). The contribution of seeds of pioneer trees buried in the soil to the regeneration of disturbed forest sites has rarely been considered (see Tiebel et al. 2018), even though the seed of pioneer tree species may make up part of soil seed banks. At different sites Augusto et al. (2001) determined birch seed densities in the soil of 45 n m^{-2}, Berger et al. (2004) found 1,039 n m^{-2}, while Mitchell et al. (1998) observed between 85-2,534 n m^{-2}. The genus *Salix* is known for its very short-lived seeds (Argus 2006; Worrell 1995), yet Staaf et al. (1987) found up to 14 n m^{-2} willow seeds in mineral soil and Bakker et al. (1996) 80 n m^{-2}. Tiebel et al. (2018) determined that there exists a general lack of information about the occurrence of rowan (*Sorbus aucuparia*) seeds in soil seed banks. Erlbeck (1998) and Hill (1979) espoused the opinion that rowan seeds remain viable up to five years in the soil due to embryo and seed coat dormancy (Raspé et al. 2000). The pioneer tree species most frequently detected in soil seed bank studies is birch (*Betula* spp.) (see Dougall and Dodd 1997; Huopalainen et al. 2001; Kjellsson 1992; Olano et al. 2002; Sullivan and Ellison 2006; Tiebel et al. 2018). Assumptions about the duration of storage in the soil of birch seeds vary from less than one year to more than 13 years (Buckley et al. 1997; Skoglund and Verwijst 1989). The findings in relation to birch seed densities in soil also varied vastly between studies depending on the number of seed sources (Tiebel et al. 2018), but this has not yet been considered in any study of soil seed banks. As part of the research carried out in the study presented here, therefore, the focus was on the storability in a temperate forest of the seed of silver birch compared to that of rowan seed.

The following five hypotheses were formulated at the outset of the study: 1) The seeds of birch and rowan may be part of the short-term persistent soil seed bank (viable 1-5 years), with a large proportion of the seeds still viable after one year (Bakker et al. 1996; Erlbeck 1998; Hill 1979; Skoglund and Verwijst 1989). 2) Rowan seeds remain viable in the soil for longer than birch (Raspé et al. 2000). 3) Seed survival rates are higher in deeper soil layers than in the upper layers (Granström and Fries 1985; Skoglund and Verwijst 1989) due to better protection against drought, light and predation (Grime et al. 1981; Tiebel et al. 2018). 4) A higher density of birch seeds in the soil seed bank is due to higher numbers of seed trees, with the result that a soil seed bank with a sufficiently high potential for regeneration after disturbance can only develop in pure and mixed birch forests. 5) The number of buried viable birch seeds decreases with increasing soil depth, because seed movement into deeper soil layers requires time and many seeds lose their viability before this movement occurs (see van Tooren 1988).

6.3 Materials and methods

The study was divided into two sub-studies: A and B. An artificial seed burial experiment (study A) was performed to study the germination capacity of *Betula pendula* L. and *Sorbus aucuparia* L. seeds stored in soil over 2.5 years. Seed burial experiments are a good means of testing the remaining viability of seeds stored in soil as the storage time, depth and the initial germination capacity of the buried seeds are all known. The aim was to determine which seeds remain viable in the soil for longer: birch (according to Tiebel et al. 2018) or rowan (according to Hill 1979). The aim of the second investigation (study B) was to determine the density of birch seeds in humus and mineral soil layers in temperate woodlands as a function of the number of available seed trees.

6.3.1 Study areas

The studies (A and B) were located at high altitudes and along the ridges of the Thuringian Forest (50°40'N and 10°45'E) and at the colline and submontane altitudes of the Tharandter Forest (50°57'N and 13°30'E) in the German Federal States Thuringia and Saxony, respectively (Burse et al. 1997; Gauer and Aldinger 2005; Fiedler and Hofmann 1978).

The area in the Thuringian Forest [*Thüringer Wald*] is situated between 400-982 m above sea level (a.s.l.), with a prevailing south-westerly exposition, many slopes and an absence of plateaus (Burse et al. 1997; Gauer and Aldinger 2005; Waesch 2003). The mean annual precipitation ranges from 700 mm in the north-east up to 1,200 mm along the ridges. The annual average temperature in the region varies between 4-6 °C. The area is influenced by an Atlantic, moderately cool and moist central mountain climate (Burse et al. 1997; Bushart and Suck 2008; Gauer and Aldinger 2005). The dominant soil types of the forest sites are low-base cambisols with low to medium nutrient contents (Gauer and Aldinger 2005).

The study area in Saxony is situated between 200-460 m a.s.l., on the northern edge of the eastern Ore Mountains [*Erzgebirge*], where it forms the transition between hill country and low mountain ranges (Fiedler and Hofmann 1978; Nebe 1982). The annual average temperature in the region varies between 7.3-7.7 °C and the mean annual precipitation ranges from 819-850 mm (Goldberg et al. 2002). The area is located at the transition between the oceanic influenced climate of the central German mountain and hill country and the continental inland climate (Nebe 1982), quite similar to the Thuringian Forest. The geology of the region has given rise to medium to deep brown soils that predominate on the forest sites, as well as dry sands and podsols with low nutrient contents and silty brown earths (Nebe 1982; Schwanecke and Kopp 1996).

Both landscapes feature a largely contiguous forest system, with ~85 % forest cover in the Saxony study area and ~90 % in the Thuringian. These are mainly single-layered, even aged Norway spruce forests (*Picea abies* (L.) Karst.). The predominant potential natural vegetation types are *Luzulo-Fagetum* and *Asperulo-Fagetum* beech forests in the Thuringian Forest and *Luzulo-Fagetum* and *Galio ordorati-Fagetum* beech forests in the Tharandter Forest (Frischbier et al. 2014; Menzer et al. 2010).

For study A, the artificial seed burial experiment, an old, pure coniferous forest in the Tharandter Forest with a closed canopy and no ground vegetation (only litter) was chosen.

For study B, soil core sampling in the forest, were taken from four different stand types differentiated by the species composition of the canopy (= stand types). Sampling was repeated three times on comparable sites (= study sites). The different stand types chosen in the Tharandter Forest were pure birch stands (Bi) and spruce stands with a small number of admixed birch trees (Bi-Sp). In the latter stand type a birch seed tree was present approximately every 50 m, corresponding to a density of 6-9 birch seed trees per hectare. The stand types considered in the Thuringian Forest consisted of spruce stands with one isolated birch tree within a radius of more than 200 m (Sp(Bi)) and pure spruce stands (Sp) (Table 6.1). The sampling sites associated with each stand type were characterized by similar soil conditions (moist with low to moderate nutrient contents) and topography to ensure that the differences in the birch seed banks in the soil were due to the contrasting stand compositions.

Table 6.1 Forest data for the sites used in the soil seed bank investigation carried out as part of study B (Bi - birch stand, Sp-Bi - mixed spruce-birch stand, Sp(Bi) - spruce stand with one birch tree, Sp - spruce stand).

location	Tharandter Forest						Thuringian Forest					
	Bi			Sp-Bi			Sp(Bi)			Sp		
study site no.	1	2	3	4	5	6	7	8	9	10	11	12
stand type	'birch'			'spruce with ad-mixed birch'			'single birch tree in spruce'			'spruce'		
tree species	*Betula pendula*			*Picea abies, Betula pendula*			*Picea abies, Betula pendula*			*Picea abies, Sorbus aucuparia*		
age of dominant tree species [year]	25	84	68	106	125	109	74	66	66	74	88	78
basal area [m²/ha]	15	42	19	37	40	40	24	35	37	21	31	31
sampling procedure												
date of sampling	Mar 2016			Mar 2016			Oct 2015			Oct 2015		
number of samples [n]	9			9			16			16		

6.3.2 Data collection

Study A - Artificial seed burial experiment

On 25 September 2015 seeds and fruits collected from mature birches (*Betula pendula* L.) and rowans (*Sorbus aucuparia* L.) (= seed sets) were filled in 10 cm x 20 cm net bags with soil and sewn up. Each bag contained either 50 birch seeds, 50 rowan seeds or 18 rowan fruits (approx. 50 rowan seeds). While removing rowan seeds from the pulp for the purposes of the experiment, it was possible to obtain an average seed number per fruit and so to calculate the number of fruits required per bag to give 50 seeds. The bags were buried at depths of 2 cm, 5 cm and 10 cm in mineral soil at two plots in a coniferous stand located 8 m apart. Two separate plots were chosen to spread the risk of loss, for example, due to seed predation or plot destruction. After burial of the bags, the original humus and litter layer were restored.

At intervals of six months between April 2016 and April 2018 sample sets (one bag per seed set, layer and plot) were removed from the soil to collect 100 seeds from each seed set and layer to test the germination rates of stored seeds. The 'seedling emergence method' was used in the greenhouse for the rowan seed sets (described in the following section; Falińska 1999). The birch seeds were placed in a climate chamber at a constant 25 °C, 80 % relative humidity and 16 h lighting per day (ISTA 2012). In addition, 400 fresh birch and rowan seeds were used to test the initial germination capacity before burial, according to the methods of the International Seed Testing Association (ISTA 2012).

Study B - soil core sampling in the forest

Seed bank sampling took place in either October or March. The samples in the Tharandter Forest were collected in March 2016, after snow-melt and before germination (Table 6.1). In the Thuringian Forest the soil samples were taken in October 2015 because the area often has a high degree of snow cover until spring. On each study site in the Thuringian Forest line transects of 150 m were established and 16 soil cores were collected at defined intervals of 1-30 m (see Huth 2009). In the Sp(Bi) study sites the transects started at the birch seed trees. Given the small size of the birch forest area (0.2-0.3 ha) in the Tharandter Forest, on each study site nine soil core samples were taken in a regular grid of 5 m x 5 m.

All cylindrical soil core samples had a diameter of 10.2 cm (81.92 cm²) and reached a depth of 10 cm into the mineral soil. The soil core samples were subdivided into three layers (= soil samples): (i) humus and litter, (ii) upper mineral soil layer (0-5 cm) and (iii) lower mineral soil layer (5-10 cm). The soil samples were stored dry for one week in the laboratory. A sieve with a mesh size of 3 mm x 4 mm was used to remove root fragments and stones from the soil samples (Gross 1990; Olano et al. 2002). The samples were filled into trays, applied in a layer of 3 cm fill height. The trays were placed in a greenhouse exposed to daylight and kept con-tinuously moist through regular watering. Additional control trays with sterilized sand were used to check for subsequent seed input in the greenhouse (contamination) over the duration of the study period. Each week the number of successfully germinated seeds was recorded and these then removed from the trays. This means of determining the number of viable seeds in the soil is referred to as the 'seedling emergence method' (Falińska 1999). The genera or spe-cies of the emerged seedlings were determined only for tree species. If the number of germi-nating seeds stagnated, the soil samples were allowed to dry, were mixed and watered again. The investigation of each sample ended 1.5 years after the date of soil collection.

6.3.3. Statistical analysis

In study A the relationship between the germination capacity of seed sets buried in soil and the factors storage time, burial depth and plot location were analyzed using logistic regres-sion. The logistic regression represented a special case of generalized linear models (GLM) due to the binominal character of the dependent variable (Faraway 2006; Zuur et al. 2009). Using the binomial GLM for either the absence or the successful germination of seeds, the expected trend towards declining seed viability with increasing storage time was modeled.

The numbers of emerged seedlings per soil sample of 81.92 cm² in study B were converted to density per m² to render the results comparable with the findings of other studies. Differences in the seed densities between study sites and stand types were analyzed using the Kruskal-Wallis H-test as the data were not normally distributed (Zar 2010). The generalized linear mixed models (GLMM) were used to assess the effect of stand type and soil layer (fixed effects) on germinated birch seeds (dependent variable). We tested whether the number of viable birch seeds in the soil depends on the number of seed tree sources and on the depth of storage in the soil. Nested random effects were the locality, study sites and sample numbers. To model the GLMM in R software version 3.3.2 (R Core Team 2014), the glmmADMB package (version 0.8.3.3.) was applied. glmmADMB used the automatic differentiation model builder (ADMB) to fit the parameters (Bolker et al. 2012). The advantages of ADMB are the range of distribution families, the range of link functions and the use of the MCMC method (Markov chain Monte Carlo) to summarize uncertainties (Bolker et al. 2008, 2012; Zuur et al. 2009). As the germinated birch seeds exhibited a negative binomial distribution, a logarithmic link function was used for modeling (Zuur et al. 2009).

Significant differences in all models were accepted at a *p-value* level of < 0.05. The necessary homoscedasticity of variance and normality were checked and confirmed with plots of residuals and quantiles from fitted models of GLMM and GLM.

6.4 Results

6.4.1 Study A - Artificial seed burial experiment

Germination rates

The initial germination capacity of freshly harvested birch seeds up to October 2015 before burial was 32 %, compared to 64 % for rowan. After the first winter period in soil, the total germination capacity of buried rowan seeds reached 82-100 %, which clearly exceeded the initial germination capacity of fresh seeds and was not observed in any other seed sets over the whole observation period. Over the rest of the storage time in soil, the germination capacity of rowan seeds and the other buried seed sets exhibited a high degree of variation. Fig. 6.1a reveals the development of the germination percentages of all seed sets in all soil layers over the study period. Rowan seeds with and without pulp had almost lost their viability in all soil layers after two years, whereas after 2.5 years the germination capacity of birch seeds was as high as at the beginning of the experiment. The germination capacity of birch seeds was always higher in spring than in autumn. The same trend was not observed for rowan. Nevertheless, the logistic regression model results showed a significant negative influence of storage time on all soil stored seed sets (Table 6.2 and Fig. 6.1b). The model results predicted that almost all rowan seeds, without and with pulp, and all birch seeds would have completely lost viability after 3, 4.5 and 12 years, respectively.

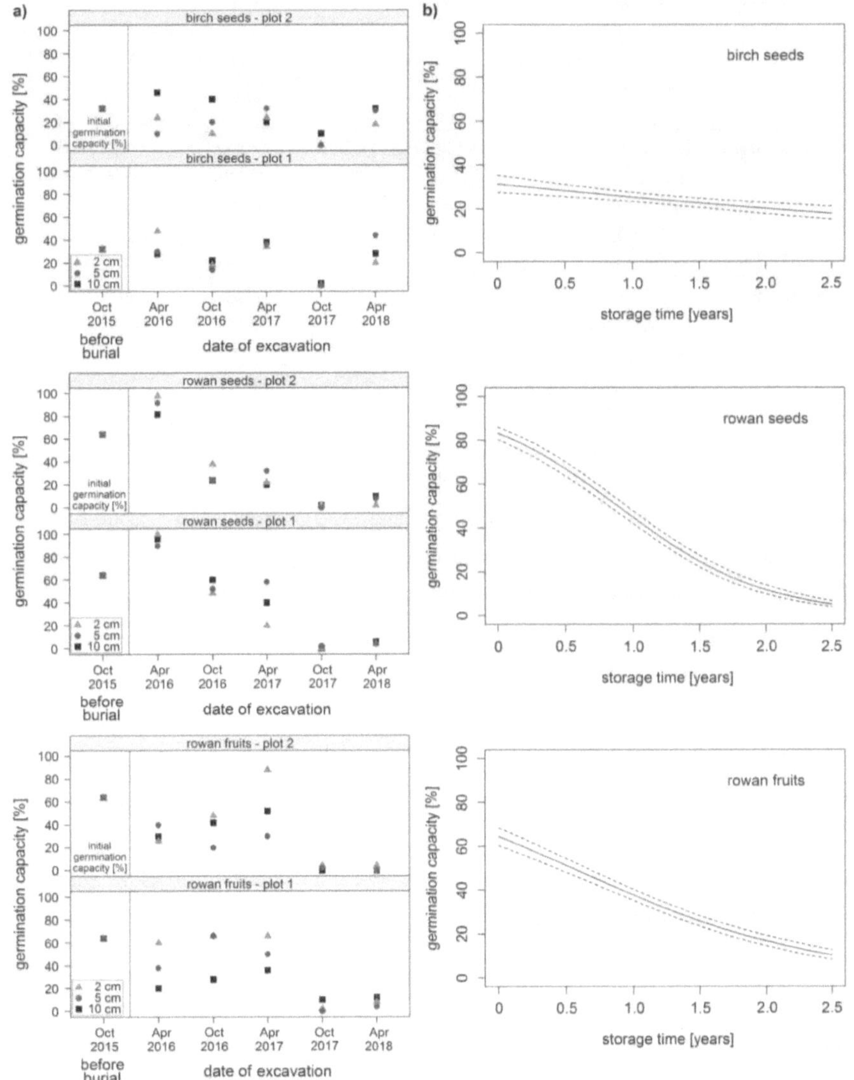

Fig. 6.1 a Total germination capacity [%] of birch seeds (top), rowan seeds (middle) and rowan fruits (bottom) stored in mineral soil at depths of 2, 5 and 10 cm in a coniferous forest. The germination capacity was tested at intervals of 6 months from April 2016 to April 2018. Mature, fleshy seed sets were buried in October 2015 before testing the initial germination capacity [%]. **b** GLM predictions of germination capacities for birch seeds (top), rowan seeds (middle) and rowan fruits (bottom) in soil over the 2.5-year study period.

Table 6.2 Logistic regression model results (GLM) for total germination capacity of buried seed sets, stored over 2.5 years in mineral soil at depths of 2, 5 and 10 cm in two study plots in a coniferous forest (n.s. - not significant).

seed sets	effects	estimate	std. error	z-value	p-value	
birch seeds	intercept	-0.863	0.135	-6.380	*0.000*	***
	storage time	-0.148	0.033	-4.516	*0.000*	***
	soil depth: 5 cm	0.097	0.139	0.695	0.487	n.s.
	soil depth: 10 cm	0.320	0.136	2.357	0.018	*
	plot 2	-0.142	0.111	-1.275	0.202	n.s.
rowan seeds	intercept	1.785	0.151	11.822	*0.000*	***
	storage time	-0.916	0.043	-21.139	*0.000*	***
	soil depth: 5 cm	0.149	0.146	1.021	0.307	n.s.
	soil depth: 10 cm	0.032	0.146	0.220	0.826	n.s.
	plot 2	-0.440	0.120	-3.669	*0.000*	***
rowan fruits	intercept	1.113	0.134	8.312	*0.000*	***
	storage time	-0.562	0.035	-15.904	*0.000*	***
	soil depth: 5 cm	-0.531	0.133	-3.994	*0.000*	***
	soil depth: 10 cm	-0.624	0.134	-4.659	*0.000*	***
	plot 2	-0.239	0.109	-2.181	*0.029*	*

The effect of storage depth on the seed viability differed between seed sets. While the germination capacity of rowan fruits decreased significantly with increasing soil depth, the model results showed no significant differences in germination capacity for rowan seeds without pulp. In contrast, birch seeds showed a significantly better germination capacity at 10 cm soil depth than in the upper two soil layers. Table 6.2 illustrates the pronounced effects of storage at the different soil depths on germination capacities for all seed sets and shows that the trend towards decreasing germination capacity over time mentioned above remains unaffected by burial depth.

The germination capacity of rowan seed sets was found to differ significantly between the two plots in the coniferous stand (Table 6.2). There was no significant difference in the case of birch. However, over time fewer seeds remained viable in the moister ground of plot two for all seed sets than in the drier soil of plot one.

Seed germination in soil

During the excavations, already germinated seeds in soil were observed for both artificial buried rowan seed sets, but not for buried birch seeds. Between 3 % and 22 % of all rowan seeds in the buried net bags germinated in spring before excavation, whereas none germinated in autumn (Fig. 6.2a top). The highest germination frequencies in soil in both plots occurred in the first spring after burial and were exhibited by the seeds buried at 2 cm (18 % and 26 %). After 2.5 years of storage the germination percentages at 2 cm had decreased to 2-4 %. However, the germination percentages at 10 cm soil depth remained similar over time. This contrasting seed germination behavior in soil was confirmed by significant differences revealed in the GLM results (Table 6.3). Nevertheless, the model results showed a generally negative effect of storage time on seed germination in the soil before excavation (Fig. 6.2b top).

In the case of the rowan fruits germination frequencies in the soil of 4-48 % were observed in the second spring. The highest germination percentages always occurred at 2 cm soil depth, after the pulp had begun to decompose (Fig. 6.2a bottom). The rowan fruits at depths of 5 cm and 10 cm exhibited no morphological changes after the first winter, whereas at 2 cm the pulp was soft and had started decomposing. By autumn 2016 no fruits were detected at 2 cm while in the lower layer the pericarps were still hard and intact. In the second autumn (2017) no fruits were found in any layer. The model results revealed significantly lower seed germinations for rowan fruits in the deeper soil layers during storage, but a significantly increasing number of germinated seeds in all layers generally over time (Table 6.3 and Fig. 6.2b bottom). The comparison, therefore, revealed that the germination behavior of the two rowan seed sets in soil over time was different.

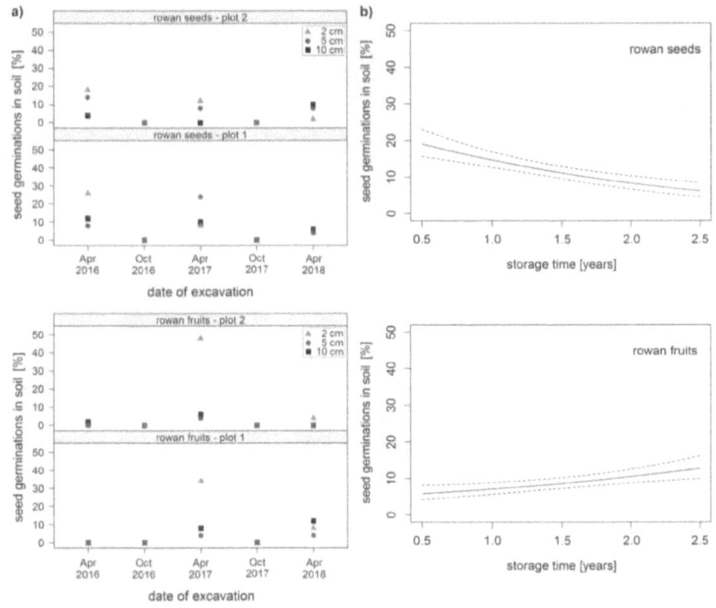

Fig. 6.2 a Seed germinations in soil [%] of total seeds buried in mineral soil at depths of 2, 5 and 10 cm for rowan seeds (top) and rowan fruits (bottom) before excavation. The early seed germination in soil was checked at intervals of 6 months from April 2016 to April 2018. Mature rowan seeds and fleshy fruits were buried in October 2015. **b** GLM predictions of seed germinations in soil for rowan seeds (top) and rowan fruits (bottom) over 2.5 years of storage.

Table 6.3 Logistic regression model results (GLM) for seed germinations in soil of buried rowan seed sets before excavation, stored over 2.5 years in mineral soil at depths of 2, 5 and 10 cm in two study plots in a coniferous forest (n.s. - not significant).

seed sets	effects	estimate	std. error	z-value	p-value	
rowan seeds	intercept	-0.766	0.215	-3.571	0.000	***
	storage time	-0.320	0.060	-5.345	0.000	***
	soil depth: 5 cm	-0.070	0.187	-0.374	0.708	n.s.
	soil depth: 10 cm	-0.587	0.209	-2.812	0.005	**
	plot 2	-0.342	0.163	-2.097	0.036	*
rowan fruits	intercept	-2.139	0.267	-8.005	0.000	***
	storage time	0.232	0.068	3.404	0.001	***
	soil depth: 5 cm	-2.260	0.315	-7.183	0.000	***
	soil depth: 10 cm	-1.376	0.227	-6.068	0.000	***
	plot 2	-0.106	0.188	-0.564	0.573	n.s.

6.4.2 Study B - Soil core sampling in the forest

Birch seedlings emerged from soil samples taken beneath all stand types, even in one sample from a pure spruce study site (Sp) (Fig. 6.3). Over the 18 month study period a total of 41,114 n m^{-2} viable birch seeds germinated in all trays.

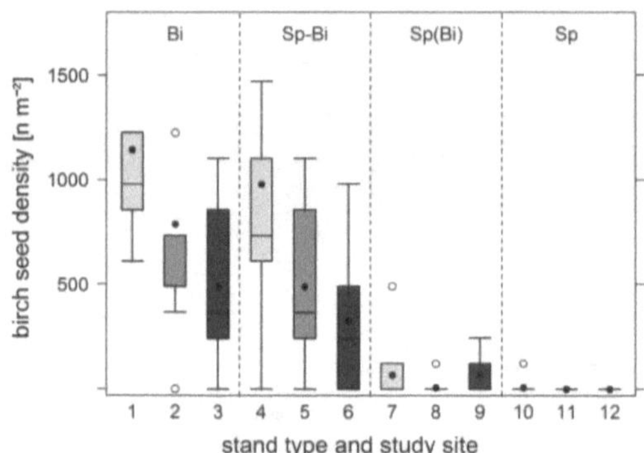

Fig. 6.3 Boxplot showing viable birch seed densities [n m^{-2}] in the soil seed banks (litter and mineral soil layers) taken from different stand types. Black lines indicate the medians, white circles indicate outliers and black circles indicate mean values (Bi - birch stand, Sp-Bi - spruce-birch stand, Sp(Bi) - spruce stand with a single birch tree, Sp - spruce stand).

Soil cores sampled from a birch stand (Bi) contained the highest mean densities of viable birch seeds (489-1,142 n m^{-2}), while samples from a spruce stand (Sp) contained the lowest mean seed densities (0-8 n m^{-2}) (Table 6.4). An average of 326-979 n m^{-2} seeds germinated from the Sp-Bi soil cores and 8-69 n m^{-2} from the Sp(Bi). The differences in the observed seed densities were significant only for Sp, Sp(Bi) and Bi (GLMM: $p = 0.000$ - Table 6.5), but not between Bi and Sp-Bi (GLMM: p-value = 0.270). The birch seed densities revealed no significant differences across the three repetitions of each stand types (Kruskal-Wallis H-test: p-value > 0.05).

Table 6.4 Viable birch seeds present in the seed banks of different stand types and soil layers, with median seed densities [n m^{-2} - in italics] and mean seed densities [n m^{-2}].

stand types	study sites	humus & litter		0-5 cm		5-10 cm		$\sum$ soil layer	
Bi birch stand	1	*612*	626	*367*	408	*0*	109	*979*	1,142
	2	*245*	394	*245*	326	*0*	68	*490*	789
	3	*367*	408	*0*	82	*0*	0	*367*	489
Sp-Bi spruce stand with admixed birch	4	*612*	653	*122*	326	*0*	0	*734*	979
	5	*367*	408	*0*	82	*0*	0	*367*	489
	6	*122*	150	*122*	136	*0*	41	*244*	326
Sp(Bi) spruce stand with a single birch tree	7	*0*	38	*0*	23	*0*	8	*0*	69
	8	*0*	0	*0*	8	*0*	0	*0*	8
	9	*0*	46	*0*	23	*0*	0	*0*	69
Sp spruce stand	10	*0*	8	*0*	0	*0*	0	*0*	8
	11	*0*	0	*0*	0	*0*	0	*0*	0
	12	*0*	0	*0*	0	*0*	0	*0*	0

Table 6.5 GLMM results for viable birch seeds in the soil samples from the different stand types and for the soil layers (f - fixed effects, r - random effects, n.s. - not significant, sd – standard deviation, study site - site number, soil core - number of the soil core from a study site). The reference stand type (intercept) is the birch stand Bi.

factor	effects	estimate	std. error	z-value	p-value		variance	sd
f	intercept	1.270	0.273	4.65	*0.000*	***		
f	stand type 'Sp-Bi'	-0.412	0.372	-1.11	0.270	n.s.		
f	stand type 'Sp(Bi)'	-2.921	0.427	-6.84	*0.000*	***		
f	stand type 'Sp'	-5.825	1.060	-5.49	*0.000*	***		
f	soil layer 0-5 cm	-0.727	0.161	-4.51	*0.000*	***		
f	soil layer 5-10 cm	-2.577	0.279	-9.24	*0.000*	***		
r	study site						0.120	0.346
r	soil core						0.412	0.642

The density of viable birch seeds decreased with increasing soil depth across all stand types (Fig. 6.4); the differences were significant (GLMM: p-value = 0.000 - Table 6.5). Birch seeds were always present in high numbers in the litter and humus layers. The density ranged from a mean seed number of 8 n m^{-2} in the spruce stand (Sp) to 626 n m^{-2} in the birch stand (Bi) (Table 6.4). No, or only very few, viable birch seeds were usually detected in lower mineral soil layers, except in the birch stand (Bi), where 68 n m^{-2} and 109 n m^{-2} seeds occurred up to 10 cm soil depth.

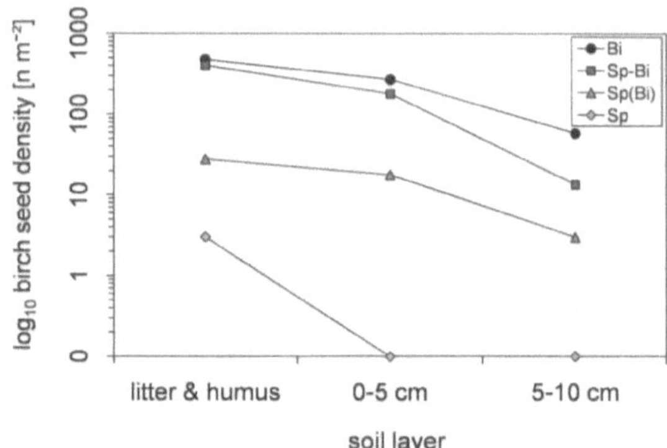

Fig. 6.4 Mean densities [n m⁻²] of viable birch seeds present in the seed bank at different soil depths for the stand types (Bi - birch stand, Sp-Bi - spruce-birch stand, Sp(Bi) - spruce stand with a single birch tree, Sp - spruce stand). Note the logarithmic scale on the y-axis.

6.5 Discussion

6.5.1 Study A - Artificial seed burial experiment

The findings of the artificial buried seed experiment showed the ability of rowan and birch seeds to remain viable for more than one year, as assumed in the first hypothesis. However, the results also confirmed contrasting storage behavior of the seed sets related to the tree species, which is discussed separately below.

Storability of birch seeds in the soil

The few buried seed experiments that have been carried out for tree species of temperate forests (Granström and Fries 1985; Granström 1987; Skoglund and Verwijst 1989) showed a viability of at least three years for *B. pendula* and *B. pubescens* seeds under the humus layer in a coniferous forest. The germination capacities in the experiments detailed by Granström and Fries (1985) and Granström (1987) saw reductions of 45 % and 94 % relative to the original germination capacity, respectively. These reductions were higher than the 6 % to 41 % observed in this study. The findings presented here correspond to those of Skoglund and Verwijst (1989), who documented 80 % viability for *B. pubescens* seeds buried in mineral soil at a depth of 10 cm in a spruce-pine forest after four years. The aforementioned authors derived a theoretical seed half-life of 13 years. Our model predicted for the same soil layer conditions a half-life of 12 years.

Skoglund and Verwijst (1989) also tested the storability of birch seeds at 10 cm soil depth on a wet meadow and found that the seeds died within one year. However, comparable with our findings, Granström and Fries (1985) found no significant differences in terms of viability loss during storage on moist sites. We assume that rapid viability loss and decomposition of birch seeds requires high levels of soil moisture, higher than occurs on moderately moist forest sites.

Tiebel et al. (2018) emphasized the importance of the beneficial effect of litter cover and thickness for higher seed storage capacity in soil. Litter protects seeds against, for example, predation and drought, and prevents early germination by light-demanding seeds (Atkinson 1992; Grime et al. 1981; Perala and Alm 1990), as we observed for birch seeds.

Bekker et al. (2000) classed the persistence of birch seeds in the soil seed bank as transient (viable < 1 year) whereas Thompson et al. (1997) classed birch seeds transient to short-term persistent (viable 1-5 years). The declining germination capacities of birch seeds observed over time in this study illustrated a slower decrease in viability, but differences in the viability rates at different soil depths make an assessment of seed persistence difficult. As per the third hypothesis, the findings suggested longer storability periods for birch in lower soil layers. Depending on the soil layer, we conclude that birch belongs to at least the short-term persistent type.

Storability of rowan seeds in the soil

Rowan seeds do not remain viable in the soil longer than birch seeds, refuting the second hypothesis. After two years most of the buried rowan seed sets with and without pulp showed very low germination capacities (3-13 %), unlike birch. Granström (1987) found 60 % of rowan seeds to be viable after three years, but only one seed germinated after five years. The declining germination capacity in Granström's study largely reflected our findings. Rowan seeds also lose their viability faster in a moist storage medium (Holmes and Buszewicz 1958, cited in Hong et al. 1996). This too was confirmed by our findings, represented by the differences in the germination success of rowan seeds between the two plots (Table 6.2 and Fig. 6.1). However, according to Erlbeck (1998) and Hill (1979), rowan seeds should be able to remain viable in the soil for up to five years (Hill 1979; Raspé et al. 2000), which we could not confirm.

The reduced storage capacity of rowan seeds in the artificial burial experiment can be explained by early seed germinations in soil before excavations (see Bradbeer 1988). The early germination in soil of rowan seeds without pulp took place in both springs (2017 and 2018), while germination of the rowan seeds with pulp only started after the second winter. Granström (1987) observed the same for buried rowan seeds. The chilling of rowan seeds under varying moisture and temperature conditions accelerates the breaking of dormancy and increases seedling emergence (Afroze and Reilly 2015; Barclay and Crawford 1984; Devillez 1979, cited in Raspé et al. 2000). Under natural stratification conditions, the dormancy of rowan seeds without pulp can be broken after 24-28 weeks (Afroze and Reilly 2015; Barclay and Crawford 1984). This corresponded to the month of April in our study. A substantial germination of rowan seeds can already take place during cold temperatures, from 2 °C (Barclay and Crawford 1984) and 5 °C (Grime et al. 1981). Devillez (1979, cited in Raspé et al. 2000) demonstrated that germination of rowan seeds not only occurs under bright conditions but also in darkness.

The successful germination of rowan seeds within fruits required longer periods of cold treatment (Barclay and Crawford 1984). First the fruit-induced secondary dormancy must be broken by decomposition of the pericarp (Bewley 1997; Devillez 1979, cited in Raspé et al. 2000). Observations made during the excavation of the net bags from the soil revealed that the decomposition of the rowan pulp starts later and proceeds more slowly deeper in the soil. Sometimes the exocarps were not destroyed in the lower soil layer, with the pulp having become dry and hard, and remaining wrapped around the seeds. This dry pulp prevented water intake and respiration by the seeds (see Bewley 1997; Perala and Alm 1990; Raspé et al. 2000). As a result, germination was inhibited and the seeds lost their viability over time. Usually, however, rowan seeds in pulp do not drift to these lower soil depths due to fruit size (see Burmeier et al. 2010; van Tooren 1988).

The findings of the artificial seed burial experiment suggest that rowan seed does not store in the soil as well as birch. However, the classification of rowan as transient (viable < 1 year) by Dölle and Schmidt (2009) and Grime et al. 1988 (cited in Raspe et al. 2000) appears incorrect. Rowan seeds should be classed in the short-term persistent soil seed bank type (seeds are viable 1-5 years in soil; Thompson et al. 1997, 1998). Nevertheless, the results indicate that an annual supply of rowan seed is required to maintain populations given their short duration in the soil seed bank.

6.5.2 Study B - Soil core sampling in the forest

Effect of forest stand types

As per the fourth hypothesis, the observed densities of viable birch seeds in the soil of 326-1,142 n m^{-2} were high in birch stands and in spruce stands with admixed birch. The birch seed density in the soil was significantly lower where there were fewer seed trees. A corresponding finding was presented by Tiebel et al. (2018). The observed densities of viable birch seed were within the range of the 224-3,760 n m^{-2} recorded in the litter and mineral soil in previous studies carried out in birch stands (Hester et al. 1991; Kjellsson1992; Miller and Cummins 2003). The seed densities of 8-69 n m^{-2} found in the vicinity of the single birch trees in the pure spruce sites were relatively low compared to the findings of Decocq et al. (2004), Dougall and Dodd (1997), Jankowska-Błaszczuk (1998), Jaroszewicz (2013) and Komulainen et al. (1994), who found 13-217 birch seeds per m^2. It seems probable that the low densities in our study arose from the fact that only one single seed tree occurred within a distance of up to 200 m. According to Ebrecht and Schmidt (2008), continuous replenishment of birch seeds by several trees is necessary to compensate the loss of viable seeds in the soil. With one exception, no birch seeds were detected in the soil beneath pure spruce stands. This absence of birch was also observed by Amrein et al. (2005) and Granström (1982).The significant differences in the numbers of birch seedlings that emerged in the trays confirmed the importance of the quantity of seed sources. This suggested birch seeds are not long-term viable, as species with long-term persistent seed banks accumulate high seed densities in the soil independent of the seed source numbers (Kjellsson 1992).

Sarvas (1948) stated that 100-200 n m^{-2} birch seeds are necessary to obtain sufficiently dense pioneer trees for the regeneration of disturbed forest sites. This implies that the soil seed banks of the spruce-birch mixed stands considered in this study, with 326, 489 and 979 n m^{-2} viable birch seeds, are sufficient for regeneration in the event of disturbance. Therefore, if an entire soil seed bank is activated after a disturbance on a site that formerly hosted spruce with about six or more admixed birch seed trees per hectare (comparable to Bi-Sp), additional birch seed rain from outside is probably not necessary for regeneration. Heinrichs (2010) and Hill (1979), on the other hand, both determined that the regeneration of birch is mainly through seed spread by seed rain and not from soil seed banks. Where there are only single seed trees or seed trees are completely absent, for example, in intact, closed spruce forest, natural regeneration after disturbance will prove difficult. A sufficient reserve of birch seed in the soil cannot be built up without seed trees and, as a consequence, the natural regeneration of these sites will depend on seed rain from surrounding forests.

Effect of soil depth

A significant decrease in the number of viable birch seeds was observed with increasing soil depth, as reported in numerous other studies (5th hypothesis; see Bakker et al. 1996; Godefroid et al. 2006; Granström 1982, 1988; Hill and Stevens 1981; Jaroszewicz 2013; Kalamees and Zobel 1998; Kjellsson 1992; Staaf et al. 1987). Granström (1988), Hill (1979) and Houle

(1998) found that the highest birch seed densities were generally in the humus layer and top-soil, representing the input of the most recent seed rain (Bakker et al. 1996; Heinrichs 2010; Houle 1998). The density of birch seeds buried under the litter and humus layers in the mineral soil is usually lower and represents an older constituent of the soil seed bank (Kjellsson 1992; Thompson et al. 1997).

During vertical drift through the soil, at an average rate of 1 cm per 6 months (Burmeier et al. 2010; Chambers and MacMahon 1994; Tiebel et al. 2018; van Tooren 1988), seeds are subjected to various causes of mortality (Holm 1994; Sarvas 1952; van Tooren 1988). In spite of the losses to mortality, long-term persistent species are known for accumulating large amounts of seed in deep mineral soil layers (Kjellsson 1992). Mortality affects the seed reserves of short-lived seeds especially severely, rapidly reducing the seed quantities in the lower soil layers (Chambers and MacMahon 1994; Kjellsson 1992). This was the case for the birch seed in this study.

6.6 Conclusions

Forest regeneration from soil seed banks can become important in situations where stands are destroyed by large scale disturbance events such as strong winds or fire. Where disturbance events affect very large areas not only are seed trees missing directly on the site but regeneration purely by means of seed rain may fail.

The artificial seed burial experiment proved the ability of birch and rowan seeds to form a short-term persistent soil seed bank. Buried birch seeds can theoretically persist in the soil under optimal storage conditions for a maximum of 12 years. Birch seeds stored better and were more tolerant of different site conditions than rowan seeds. Rowan seeds, without and with pulp, germinated independent of the soil depth after the first and second winter in the soil, which limited seed storability in the soil significantly. Rowan seeds without pulp could be stored for a maximum of 3 years, and seeds with pulp 4.5 years. The pulp surrounding rowan seeds offers no benefits in terms of seed storability; rather it would appear to act as a physical inhibitor on germination and ultimately leads to reduced seed viability. Therefore, a continuous, almost annual input of rowan seeds to the soil seed bank is necessary, while birch seed input every few years seems sufficient.

The results of the soil core sampling in the forest showed a clear relationship between the numbers of seed tree sources in the stands and the densities of viable birch seed in the soil. To build up a soil seed bank sufficient for the successful regeneration of disturbed forest sites, in spruce dominated stands more than six birch seed trees are required per hectare. This corresponds to a distance of not more than 50 m between seed trees. Lower seed tree densities per hectare lead to low birch seed reserves in the soil and insufficient replenishment of seed banks with fresh seed.

6.7 References

Afroze F, O'Reilly C (2015) Effect of harvest date, drying, short-term storage and freezing after chilling on the germination of rowan seeds. Scand. J. For. Res. 31:339-346. 10.1080/02827581.2015.1080293

Amrein D, Rusterholz H-P, Baur B (2005) Disturbance of suburban *Fagus* forests by recrea-

tional activities: effects on soil characteristics, aboveground vegetation and seed bank. Appl. Veg. Sci. 8:175-182. 10.1658/1402-2001(2005)008[0175:DOSFFB]2.0.CO;2

Argus GW (2006) Guide to *Salix* (willow) in the Canadian Maritime Provinces (New Brunswick, Nova Scotia, and Prince Edward Island). Canadian Museum of Nature, Ottawa, Canada. http://accs.uaa.alaska.edu/files/botany/publications/2006/GuideSalixCanadian AtlanticMaGuideS.pdf Accessed 10 Feb 2018

Atkinson MD (1992) *Betula pendula* Roth (*B. verrucosa* Ehrh.) and *B. pubescens* Ehrh. J. Ecol. 80:837-870. 10.2307/2260870

Augusto L, Dupouey J-L, Picard J-F, Ranger J (2001) Potential contribution of the seed bank in coniferous plantations to the restoration of native deciduous forest vegetation. Acta Oecol. 22:87-98. 10.1016/S1146-609X(01)01104-3

Bakker JP, Bakker ES, Rosén E, Verwej GL, Bekker RM (1996) Soil seed bank composition along a gradient from dry alvar grassland to *Juniperus* shrubland. J Veg Sci. Science 7:165-176. 10.2307/3236316

Barclay AM, Crawford RMM (1984) Seedling emergence in the rowan (*Sorbus aucuparia*) from an altitudinal gradient. J. Ecol. 72:627-636. 10.2307/2260072

Bekker RM, Verwej GL, Bakker JP, Fresco LFM (2000) Soil seed bank dynamics in hayfield succession. J. Ecol. 88:594-607. 10.1046/j.1365-2745.2000.00485.x

Berger TW, Sun B, Glatzel G (2004) Soil seed banks of pure spruce (*Picea abies*) and adjacent mixed species stands. Plant and Soil 264:53-67. 10.1023/B:PLSO.0000047753.36424.41

Bernhardt K-G, Poschlod P (1992) Diasporenbank im Boden als Vegetationsbestandteil Teil 1: Europa. Excerpta Botanica Sec. B 29:241-260

Bewley JD (1997) Seed germination and dormancy. The Plant Cell 9:1055-1066. 10.1105/tpc.9.7.1055

Bolker BM, Brooks ME, Clark CJ, Geange SW, Poulsen JR, Stevens MHH, White J-DS (2008) Generalized linear mixed models: a practical guide for ecology and evolution. Trends Ecol Evolut 24:127-135. 10.1016/j.tree.2008.10.008

Bolker B, Skaug H, Magnusson A, Nielsen A (2012) Getting started with the glmmADMB package. http://glmmadmb.r-forge.r-project.org/glmmADMB.pdf Accessed 11 Feb 2019

Bossuyt B, Hermy M (2001) Influence of land use history on seed banks in European temperate forest ecosystems: a review. Ecography 24:225-238. 10.1034/j.1600-0587.2001.240213.x

Bossuyt B, Honnay O (2008) Can the seed bank be used for ecological restoration? An overview of seed bank characteristics in European communities. J Veg Sci. 19:875-884. 10.3170/2008-8-18462

Bradbeer JW (1988) Seed dormancy and germination. Blackie and Son Ltd, Glasgow and London

Buckley GP, Howell R, Anderson MA (1997) Vegetation succession following ride edge management in lowland plantations and woods. 2. The seed bank resource. Biol Cons 82:305-316. 10.1016/S0006-3207(97)00025-6

Burmeier S, Eckstein RL, Otte A, Donath TW (2010) Desiccation cracks act as natural seed traps in flood-meadow systems. Plant Soil 333:351-364. 10.1007/s11104-010-0350-1

Burse K-D, Schramm H-J, Geiling S, Meinhardt H, Schölch M (1997) Die forstlichen Wuchsbezirke Thüringens - Kurzbeschreibung. Mitteilungen der Landesanstalt für Wald und Forstwirtschaft, Gotha

Bushart M, Suck R (2008) Potenzielle natürliche Vegetation Thüringens, Schriftenreihe der Thüringer Landesanstalt für Umwelt und Geologie, Jena

Chambers JC, MacMahon JA (1994) A day in the life of a seed: movements and fates of seeds and their implications for natural and managed systems. Annu. Rev. Ecol. Syst 25:263-292. https://doi.org/10.1146/annurev.es.25.110194.001403

Decocq G, Valentin B, Toussaint B, Hendoux F, Saguez R, Bardat J (2004) Soil seed bank composition and diversity in a managed temperate deciduous forest. Biodivers Conserv 13:2485-2509. 10.1023/B:BIOC.0000048454.08438.c6

Dölle M, Schmidt W (2009) The relationship between soil seed bank, above-ground vegetation and disturbance intensity on old-field successional permanent plots. Appl. Veg. Sci. 12:415-428. 10.1111/j.1654-109X.2009.01036.x

Dougall TAG, Dodd JC (1997) A study of species richness and diversity in seed banks and its use for the environmental mitigation of a proposed holiday village development in a coniferized woodland in south east England. Biodiversity and Conservation 6:1413-1428. 10.1023/A:1018345915418

Ebrecht L, Schmidt W (2008) Bedeutung der Bodensamenbank und des Diasporentransports durch Forstmaschinen für die Entwicklung der Vegetation auf Rückegassen. Forstarchiv 79:91-105

Egawa C, Tsuyuzaki S (2013) The effects of litter accumulation through succession on seed bank formation for small and large seeded species. J Veg Sci. 24:1062-1073. 10.1111/jvs.12037

Erlbeck R (1998) Die Vogelbeere (*Sorbus aucuparia*) - ein Porträt des Baumes des Jahres 1997. Berichte aus der Bayrischen Landesanstalt für Wald und Forstwirtschaft 17:2-14

Falińska K (1999) Seed bank dynamics in abandoned meadows during a 20-year period in the Białowieża National Park. . J. Ecol. 87:461-475. 10.1046/j.1365-2745.1999.00364.x

Faraway JJ (2006) Generalized linear, mixed effects and nonparametric regression models. Chapman & Hall/CRC Taylor & Francis Group, London, New York

Fiedler HJ, Hofmann W (1978) Standortskundlicher Exkursionsführer „Tharandt-Grillenburger Wald" - Bodennutzung und Bodenschutz. Studienmaterial für die Weiterbildung. TU Dresden - Direktion Studienangelegenheiten, Dresden

Frischbier N, Profft I, Hagemann U (2014) Potential impacts of climate change on forest habitats in the Biosphere Reserve Vessertal-Thuringian Forest in Germany. In: Rannow S, Neubert M (eds) Managing protected areas in central and eastern Europe under climate change, advances in global change research. Springer Science and Business Media, Dordrecht, pp 243-257. https://doi.org/10.1007/978-94-007-7960-0_16

Gauer J, Aldinger E (2005) Waldökologische Naturräume Deutschlands: Forstliche Wuchsgebiete und Wuchsbezirke - mit Karte 1:1.000.000. Mitteilungen des Vereins für Forstliche Standortskunde und Forstpflanzenzüchtung, Stuttgart

Godefroid S, Phartyal SS, Koedam N (2006) Depth distribution and composition of seed banks under different tree layers in a managed temperate forest ecosystem. Acta Oecol. 29:283-292. 10.1016/j.actao.2005.11.005

Goldberg V, Baums AB, Häntzschel J (2002) Klima, Boden und Landnutzung. In: Bernhofer C, Berger FH, Goldberg V (eds): Exkursions- und Praktikumsführer Tharandter Wald. Tharandter Klimaprotokolle Band 6. Eigenverlag der Technischen Universität Dresden, Dresden, pp 15-26

Graber RE, Thompson DF (1978) Seeds in the organic layers and soil of four beech-birch-maple stands. Northeastern Forest Experiment Station, Forest Service Research Paper NE-401.U.S, Department of Agriculture, Broomall, USA

Granström A (1982) Seed banks in five boreal forest stands originating between 1810 and 1963. Can J Bot 60:1815-1821. 10.1139/b82-228

Granström A (1987) Seed viability of fourteen species during five years of storage in a forest soil. J. Ecol. 75: 321-331. 10.2307/2260421

Granström A (1988) Seed banks at six open and afforested heathland sites in southern Sweden. J Appl Ecol 25: 297-306. 10.2307/2403627

Granström A, Fries C (1985) Depletion of viable seeds of *Betula pubescens* and *Betula verrucosa* sown onto some north Swedish forest soils. Can J Forest Res. 15:1176-1180. 10.1139/x85-191

Grime JP, Mason G, Curtis AV, Rodman J, Band SR (1981) A comparative study of germination characteristics in a local flora. . J. Ecol. 69:1017-1059. 10.2307/2259651

Gross KL (1990) A Comparison of methods for estimating seed numbers in the soil. J. Ecol. 78:1079-1093. 10.2307/2260953

Heinrichs S (2010) Response of the understorey vegetation to select cutting and clear cutting in the initial phase of Norway spruce conservation. PhD book, Faculty of Mathematics and Natural Sciences, Georg-August-University Göttingen, Göttingen

Hester AJ, Gimingham CH, Miles J (1991) Succession from heather moorland to birch woodland. III. Seed availability, germination and early growth. J. Ecol. 79:329-344. 10.2307/2260716

Hill MO (1979) The development of a flora in even-aged plantations. In: Ford ED, Malcolm DC, Atterson J (eds) The ecology of even-aged forest plantations. Institute of Terrestrial Ecology, Cambridge, pp 175-192

Hill MO, Stevens PA (1981) The density of viable seed in soils of forest plantations in upland Britain. J. Ecol. 69:693-709. 10.2307/2259692

Holm SO (1994) Reproductive patterns of *Betula pendula* and *B. pubescens* coll. along a regional altitudinal gradient in northern Sweden. Ecography 17:60-72. 10.1111/j.1600-0587.1994.tb00077.x

Hong TD, Linington S, Ellis RH (1996) Seed storage behaviour: a compendium. Handbooks for Genebanks no. 4. International Plant Genetic Resources Institute, Rome

Hopfensperger KN (2007) A review of similarity between seed bank and standing vegetation across ecosystems. Oikos 116:1438-1448. 10.1111/j.2007.0030-1299.15818.x

Houle G (1998) Seed dispersal and seedling recruitment of *Betula alleghaniensis*: spatial inconsistency in time. Ecology 79:807-818. 10.1890/0012-9658(1998)079[0807:SDASRO]2.0.CO;2

Huopalainen M, Tuittila E-S, Vanha-Majamaa I (2001) Effects of long-term aerial pollution on soil seed banks in drained pine mires in southern Finland. Water Air Soil Pollut 125:69-79. https://doi.org/10.1023/A:1005276201740

Huth F (2009) Untersuchungen zur Verjüngungsökologie der Sand-Birke (*Betula pendula* Roth). PhD book, Faculty of Forest, Geo and Hydro, Dresden University of Technology, Dresden

ISTA (2012) Internationale Vorschriften für die Prüfung von Saatgut 2012. International Seed Testing Association (ISTA), Bassersdorf

Jankowska-Błaszczuk M (1998) Variability of the soil seed banks in the natural deciduous forest in the Białowieża National Park. Acta Soc Bot Pol. 67:313-324. 10.5586/asbp.1998.040

Jaroszewicz B (2013) Endozoochory by European bison influences the build-up of the soil seed bank in subcontinental coniferous forest. Eur J Forest Res 132:445-452. 10.1007/s10342-013-0683-4

Kalamees R, Zobel M (1998) Soil seed bank composition in different successional stages of a species rich wooded meadow in Laelatu, western Estonia. Acta Oecol. 19:175-180. 10.1016/S1146-609X(98)80021-0

Kjellsson G (1992) Seed banks in Danish deciduous forests: species composition, seed influx and distribution pattern in soil. Ecography 15:86-100. 10.1111/j.1600-0587.1992.tb00012.x

Komulainen M, Vieno M, Yarmishko VT, Daletskaja TD, Maznaja EA (1994) Seedling establishment from seeds and seed banks in forests under long-term pollution stress: a potential for vegetation recovery. Can J Bot 72:143-149. 10.1139/b94-019

Landenberger RE, McGraw JB (2004) Seed-bank characteristics in mixed-mesophytic forest clearcuts and edges: Does "edge effect" extend to the seed bank? Can. J. Bot. 82: 992-1000. https://doi.org/10.1139/b04-080

Leck MA, Parker VT, Simpson RL (2008) Seedling ecology and evolution. University Press, Cambridge

Leder B (1992) Weichlaubhölzer: Verjüngungsökologie, Jugendwachstum und Bedeutung in Jungbeständen der Hauptbaumarten Buche und Eiche. Landesanstalt für Forstwirtschaft Nordrhein-Westfalen, Arnsberg

Marquis DA (1975) Seed storage and germination under northern hardwood forests. Can. J. For. Res. 5:478-484. 10.1139/x75-065

Menzer A, Feger K-H, Lohse H, Joisten H, Hurst S (2010) Bodenlehrpfad Tharandter Wald - Exkursionsführer. Sächsisches Landesamt für Umwelt, Landwirtschaft und Geologie, Dresden

Miller GR, Cummins RP (2003) Soil seed banks of woodland, heathland, grassland, mire and montane communities, Cairngorm Mountains, Scotland. Plant Ecol. 168:255-266. 10.1023/A:1024464028195

Mitchell RJ, Marrs RH, Auld MHD (1998) A comparative study of the seedbanks of heathland and successional habitats in Dorset, southern England. J. Ecol. 86:588-596. 10.1046/j.1365-2745.1998.00281.x

Nebe W (1982) Natürliche Grundlagen des Waldwachstums und der Waldentwicklung. Der Tharandter Wald. Forststadt Tharandt - Beitrag zur Heimatgeschichte 7/82, Gemeindeverband Tharandt

Olano JM, Caballero I, Laskurain NA, Loidi J, Escudero A (2002) Seed bank spatial pattern in a temperate secondary forest. J Veg Sci. 13:775-784. https://doi.org/10.1111/j.1654-1103.2002.tb02107.x

Olmsted NW, Curtis JD (1947) Seeds of the forest floor. Ecology 28:49-52. 10.2307/1932917

Perala DA, Alm AA (1990) Reproductive ecology of birch: a review. Forest Ecol Manag 32:1-38. 10.1016/0378-1127(90)90104-J

R Core Team (2014) R: A language and environment for statistical computing. R Foundation for Statistical Computing, Vienna. https://www.R-project.org Accessed 11 Feb 2019

Raspé O, Findlay C, Jacquemart A-L (2000) *Sorbus aucuparia* L. J. Ecol. 88:910-930. 10.1046/j.1365-2745.2000.00502.x

Sarvas R (1948) A research on the regeneration of birch in South Finland. Commun. Inst. For. Fenn. 35:82-91

Sarvas R (1952) On the flowering of birch and the quality of seed crop. Commun. Inst. For. Fenn. 40:1-35

Schwanecke W, Kopp D (1996) Forstliche Wuchsgebiete und Wuchsbezirke im Freistaat Sachsen. Schriftenreihe der Sächsischen Landesanstalt für Forsten 8/96

Skoglund J, Verwijst T (1989) Age structure of woody species populations in relation to seed rain, germination and establishment along the river Dalälven, Sweden. Vegetation 82:25-34. 10.1007/BF00217979

Staaf H, Jonsson M, Olsén L-G (1987) Buried germinative seeds in mature beech forests with different herbaceous vegetation and soil types. Holarctic Ecol 10:268-277. 10.1111/j.1600-0587.1987.tb00768.x

Sullivan KA, Ellison AM (2006) The seed bank of hemlock forests: implications for forest regeneration following hemlock decline. J Torrey Bot Soc 133:393-402.10.3159/1095-5674(2006)133[393:TSBOHF]2.0.CO;2

Telewski FW, Zeevaart JAD (2002) The 120-yr period for Dr. Beals´s seed viability experiment. Am J Bot 89:1285-1288. 10.3732/ajb.89.8.1285

Thompson K, Bakker J, Bekker R (1997) The soil seed banks of North West Europe: methodology, density and longevity. Cambridge University Press, Cambridge

Thompson K, Bakker JP, Bekker RM, Hodgson JG (1998) Ecological correlates of seed persistence in soil in the north-west European flora. J. Ecol. 86:163-169. 10.1046/j.1365-2745.1998.00240.x

Tiebel K, Huth F, Wagner S (2018) Soil seed banks of pioneer tree species in European temperate forests: a review. iForest 11:48-57. 10.3832/ifor2400-011

Tooren BF van (1988) The fate of seeds after dispersal in chalk grassland: the role of the bryophyte layer. Oikos 53:41-48. 10.2307/3565661

Waesch G (2003) Montane Graslandvegetationen des Thüringer Waldes: Aktueller Zustand, historische Analyse und Entwicklungsmöglichkeiten, 1st ed. Cuvillier Verlag, Göttingen

Worrell R (1995) European aspen (*Populus tremula* L.): a review with particular reference to Scotland: I. Distribution, ecology and genetic variation. Forestry 68:93-105. 10.1093/forestry/68.2.93

Zar JH (2010) Biostatistical analysis, 5th ed. Prentice Hall, Upper Saddle River, New Jersey

Zerbe S (2001) On the ecology of *Sorbus aucuparia* (*Rosaceae*) with special regard to germination, establishment and growth. Pol. Bot. J. 46:229-239

Zobel M, Kalamees R, Püssa K, Roosaluste E, Moora M (2007) Soil seed bank and vegetation in mixed coniferous forest stands with different disturbance regimes. Forest Ecol Manag 250:71-76. 10.1016/j.foreco.2007.03.011

Zuur AF, Ieno EN, Walker NJ, Saveliev A, Smith GM (2009) Mixed effects and extensions in ecology with R. Springer Science and Business Media, New York

Chapter 7

General discussion

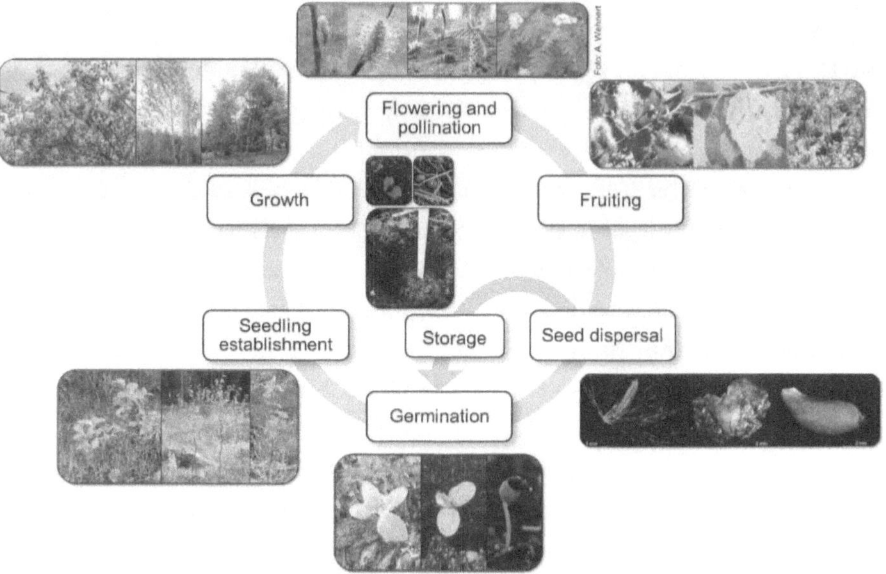

7.1 Discussion of important aspects of regeneration ability

Fructification is essential to the successful establishment of young trees, with successful regeneration the key to safeguarding the survival and conservation of species in forest ecosystems. Successful regeneration requires successful seed germination, seedling establishment and ultimately trees maturing to where they too fructify and produce seed.

Trees need microsites, also called safe sites and regeneration niches, for successful germination (Harper 1977, Abé 2002, Baier et al. 2007). These regeneration niches must fulfill the species' specific requirements in relation to environmental factors, such as light, humidity, temperature, substrate, soil moisture and competition (Röhrig et al. 2006, Bartsch & Röhrig 2016). Each tree species has different requirements of regeneration niches based on their autecology. The requirements of regeneration niches can be classified generally and analyzed according the ecological groups of tree species: shade-tolerant, intermediate and pioneer tree species.

Shade-tolerant tree species, like *Fagus sylvatica* L. and *Abies alba* Mill., need regeneration niches with constant humidity, soil moisture, well-drained microsites and low competition from herbaceous species, but do not require particularly high light intensity (Harcombe et al. 1982, Löf 2000, Bucher 2008, Bartsch & Röhrig 2016, Huth et al. 2017). Suitable regeneration niches are mainly found in closed forests and beneath small canopy gaps. Seedlings of shade-tolerant tree species can establish under a closed canopy and persist growing very slowly for many years until an opening of the canopy leads to higher light availability and then grow more quickly (Abé 2002). Pioneer tree species primarily need moist microsites with bare ground and high light intensity for fast germination and establishment (Lautenschlager 1994, Zerbe 2001, Gage & Cooper 2005, Mihók et al. 2005, Röhrig et al. 2006). Regeneration niches of pioneer tree species are limited to large gaps and open sites. Like shade-tolerant tree species, intermediate tree species, such as *Picea abies* L. and *Quercus* ssp., are vulnerable to drought and high competition from herbaceous species. Unlike shade-tolerant species, seedlings require light conditions between 20-40 % of open land conditions (Baier et al. 2007, Reif & Gärtner 2007, Volkert & Reif 2010, Bartsch & Röhrig 2016), with the result that their regeneration niches are found in middle (< 200 m²; Runkle & Yetter 1987) to large gaps (> 400 m²; de Lima et al. 2012) and along forest edges.

The aforementioned regeneration sites – closed forest, gaps, forest edges and open sites – are not equally distributed in forest ecosystems. Large gaps, including open sites, are much less frequently found in closed forest structures (< 10 % of all gaps) than small gaps. Small gaps often account for more than 50 % of all gaps recorded in forests (Runkle & Yetter 1987,

Lertzman & Krebs 1991, Liu & Hytteborn 1991, Runkle 1992, Huth & Wagner 2006, de Lima et al. 2012). These small gaps usually arise with the death of one or a small number of trees in the dominant stand layer. Less common large-scale disturbance events, like insect calamities, storm and fire, result in large gaps up to open sites. The distribution of gap sizes in a forest resembles a lognormal function (Liu & Hytteborn 1991, Huth & Wagner 2006, de Lima et al. 2012).

The frequency and distribution of the different regeneration sites in a forest ecosystem requires more or less large investments by the tree species in spatial and temporal seed distribution. For example, spatial seed distribution over long distance can be done anemochorously and zoochorously. To reach a newly created, large gap or open site with low vegetation cover by anemochorous seed dispersal requires the production of very small and light seeds without a nutrient reserve. Correspondingly, these seeds have greater requirements of their regeneration niches in terms of soil moisture and conditions facilitating fast germination (see Gage & Cooper 2005, Mihók et al. 2005, Schütt et al. 2011). If these small, light seeds are deposited on unfavorable sites, seedling mortality increases drastically and seedling establishment is hindered (Bramlett 1990, Gage & Cooper 2005, Mihók et al. 2005, Röhrig et al. 2006).

By contrast, zoochorous seed dispersal has the advantage of a more deliberate distribution of seeds by the animal vector. There is a higher probability that these seeds can reach appropriate regeneration niches than anemochorously dispersed seeds (Clark et al. 1998, Żywiec et al. 2013). However, zoochorous tree species must produce attractive, energy-rich fruits or seeds that are suitable as food for the consumer (Gautier-Hion et al. 1985, Silveira et al. 2013). The production of large fruits or seeds may result in resource depletion in the seed tree (Gurnell 1993, Sork 1993, Wohlgemuth et al. 2016), with the result that annual production of a large amount of fruits/seeds is not possible every year, unlike wind dispersed tree species (Chmelar & Meusel 1986, Atkinson 1992, Raspé et al. 2000). In the case of the heavy-seeded tree species *Fagus sylvatica* and *Quercus* ssp., resource depletion after a mast year regularly results in very low seed production in subsequent years (Sork 1993, Hilton & Packham 2003, Övergaard et al. 2007, Wohlgemuth et al. 2016).

Tree species with regeneration niches widely distributed throughout forests, for example, shade-tolerant tree species, do not need to produce small, light seeds with a high investment in their spatial distribution. These tree species can produce big and heavy seeds with a food reserve.

Temporal seed distribution can take place through either a soil seed bank or seedling bank. Tree species obtain a temporal lead, if they have seeds distributed in the soil or seedlings established under a closed canopy. The seeds or seedlings wait for improved environmental conditions, like increased irradiation following gap formation (Thompson et al. 1997, Abé 2002). Only shade-tolerant or climax tree species can build up a seedling bank, whereas a soil seed bank requires that seeds be storable, a characteristic of the seeds of pioneer species (see Brokaw 1986, Abé 2002, Hopfensperger 2007).

The various possible means of spatial and temporal seed dispersal by trees of various species begs the question whether the ecological groups of tree species (shade-tolerant, intermediate and pioneer tree species) have uniform regeneration strategies or not. The assumption of uniform strategies seems possible given the aforementioned general explanations of the regeneration niches. However, looking at the spatial distribution of seeds, the question of uniform regeneration strategies must already be answered with no. Each ecological group includes tree species with seed distributed by wind and by fauna. Shade-tolerant tree species do not exclusively produce the kind of large and heavy seeds that are supposed to be beneficial to them (e.g., *Fagus sylvatica*), but also light and small wind-dispersed seeds (e.g., *Abies alba*) (Bucher 2008, Huth et al. 2017). Examples of wind-dispersed intermediate tree species are *Tilia* ssp. and *Ulmus* ssp., while zoochorously dispersed intermediate tree species are *Quercus* ssp. and *Malus sylvestris* (Burschel & Huss 1997). The pioneer tree species include zoochorously-dispersed tree species such as *Sorbus aucuparia* in temperate forests and *Miconia argentea* and *Cecropia insignis* in tropical forests, as well as the wind-dispersed species *Salix* ssp., *Populus* ssp. and *Betula* ssp. (Brokaw 1987, Burschel & Huss 1997). Interestingly, the seeds of the wind-dispersed tree species *Pinus sylvestris* are also consumed and dispersed by 28 bird species (Schütt & Stimm 2006).

Seed morphology also differs within the ecological groups of tree species, and the corresponding possibilities for spatial seed dispersal. The shade-tolerant zoochorously-dispersed *Taxus baccata* produces small seeds of 6-7 mm x 3-5 mm in size (Schütt 2008) with a thousand-seed weight of 7-70 g (MacCarthaigh & Spethmann 2000), while *Fagus sylvatica* produces 15-20 mm x 10 mm seeds (Hecker 1998, Kandemir & Kaya 2009, Kremer 2010) with a thousand-seed weight of 220-250 g (MacCarthaigh & Spethmann 2000). Zoochorously-dispersed pioneer tree species with very small seeds and fruits are *Miconia argentea* (fruit size of 0.5-3.5 cm in diameter, Silveira et al. 2013) and *Sorbus aucuparia* (fruit size of 1.0-1.4 cm, Raspé et al. 2000). Brokaw (1987) mentioned that the tropical pioneer species *Cecropia insignis* also has very small seeds, but the fruits are many times larger (15 cm long and 1.3 cm in diameter, Smithsonian undated). In comparison to the wind-dispersed pioneer species *Salix caprea* (thousand-seed weight: < 0.1 g, Schütt & Stimm 2001), *Pinus sylvestris* –

also a pioneer species with wind-dispersed seeds – forms quite large and heavy winged seeds (thousand-seed weight: 7-17 g, MacCarthaigh & Spethmann 2000). The sinking rate of *Pinus sylvestris* seed is 1.45 sec m^{-1}, which is similar to that of the seed of the intermediate tree species *Acer platanoides* (sinking rate of 1.13 sec m^{-1}) (Kohlermann 1950).

Temporal seed distribution also differs within the ecological groups of tree species. Tropical pioneer tree species are known for their potential to build up long-term persistent soil seed banks (Dalling et al. 1998), unlike temperate pioneer tree species. However, whereas the tropical pioneer tree *Cecropia insignis* forms only a transient soil seed bank (Dalling et al. 1998), the seeds of the temperate *Pinus sylvestris* are able to remain viable for 4-5 years after ripening (Schütt & Stimm 2006, Baumann 2007). The most shade-tolerant tree species of temperate forests should in theory only be able to establish a seedling bank (Brokaw 1986, Abé 2002, Stancioiu & O'Hara 2006). *Taxus baccata* forms both a seedling bank and a seed bank (Iszkuło et al. 2005), however, and the intermediate tree species *Quercus robur* and *Q. sessisiliflora* may form a seedling bank under certain light conditions (e.g., in pine stands) (Reif & Gärtner 2007, Volkert & Reif 2010).

Ultimately it would appear that all of the described possibilities for the distribution of seed can occur in different and varied combinations (spatial and temporal). No clear regeneration strategy based on the ecological groups prevails. It may also be that not all distribution possibilities are combinable. It is necessary to have a look at each tree species, to analyze their individual regeneration strategy and to clarify their specific characteristics. Only then might it be possible to arrive at conclusions in relation to their regeneration ecology and to design corresponding silvicultural measures to promote their establishment.

As was described in chapter 1, the main goal of the study was to obtain comprehensive knowledge of the ecological aspects of the regeneration cycles of the temperate pioneer tree species *Salix caprea*, *Betula pendula* and *Sorbus aucuparia* to close existing gaps in the knowledge. However, not all aspects of the ecology and of the processes occurring within the regeneration cycle of these species (see Fig. 1.1, p. 7) could be studied (Fischer et al. 2016). The focus of the study was on fructification, seed dispersal and seed storage in soil seed banks (see chapter 1.3 and chapter 1.4, pp. 9-13). To obtain a comprehensive overview and to discuss the whole regeneration cycles of the pioneer tree species investigated, information relating to aspects of the regeneration cycle not researched has been supplemented by information drawn from the relevant literature. The objective was to provide a full picture of the possibilities and limits of the natural regeneration of disturbed sites in spruce-dominated forests at high elevations and along ridges by goat willow, silver birch and rowan.

7.1.1 Fructification and seed production in *Salix caprea, Betula pendula* and *Sorbus aucuparia*

The results showed a strong influence of weather conditions on the fructification of and seed production by the pioneer tree species studied, as assumed in the first hypothesis (p. 10). This finding was independent of the other impacting factors, namely vitality, age and tree dimensions (Moles et al. 2004, Żywiec et al. 2012, Fischer et al. 2016). The study revealed that the seed numbers produced by silver birch were four times lower under suboptimal weather conditions, and one and a half times lower in the case of goat willow. As was described and discussed in chapters 2.5.1, p. 42 ff. and chapter 3.5.1, p. 71 ff., these observations were confirmed by numerous other studies (Sarvas 1948, Bastide & van Vredenburch 1970, Elmqvist et al 1988, Houle & Payette 1990, Kullmann 1993, Holm 1994, Gage & Cooper 2005, Huth 2009). Varying seed production numbers were also observed for rowan in different studies (Prien 1964, Sperens 1997a, b, Raspé et al. 2000, Satake et al. 2004, Żywiec et al. 2012). Żywiec et al. (2012) reported that mast years in rowan are mainly influenced by weather conditions and not by predation pressure as is often assumed (Sperens 1997a). Seed production by rowan seed trees at high altitudes in the Thuringian Forest was either insufficient or failed entirely in 2015 and 2016, whereas the rowans in the lowlands were fruit bearing. The reasons for the seed crop failures were probably the dry autumns, winters and springs in both years, and the hot summer in 2015 (Thüringer Klimaagentur 2015a, b). Normally at these high altitudes precipitation is plentiful (e.g., chapter 4.3, p. 88 ff. - Burse et al. 1997, Gauer & Aldinger 2005, Bushart & Suck 2008). This stressed the individuals, reduced their fitness and impacted negatively on seed production by rowan trees (see Sperens 1997b).

Although the investigation of seed production by individual pioneer seed trees was not the main focus of this study, the quantities of seeds collected in the traps and the model results provide good indications of the differences between the pioneer tree species, supplemented by helpful information from other studies. The predicted level of seed production by a silver birch seed tree with 20 cm dbh was 1.5 million seeds in the non-mast year and 3.5 million seeds in the mast year. In mast years a single tree can produce seed crops as high as 7.3-10.0 million seeds (dbh of 24-80 cm - Arnbourg 1948, cited in Perala & Alm 1990, Popadyuk et al. 1995, Wagner et al. 2004, Huth 2009). In non-mast years birch seed production can fall to 30,000-40,000 seeds per tree (Denisow 2007, cited in Huth 2009). Adult goat willow seed trees of low to medium vitality may produce 1.0-22 million seeds (dbh of 11-37 cm) in a particular year according to Tiebel et al. (2019). The same authors found that better weather conditions and more vital individuals result in higher seed numbers. The assumption was confirmed by the finding of seed crops of 740,000 seeds produced by willow seed trees of 3 m in height (Karrenberg & Suter 2003). Rowan seed trees less than 10 m in height may produce between 1-4,322 fruits per tree in mast and non-mast years, with a recorded maximum of 22,542 fruits per tree (Sperens 1997b, Żywiec et al. 2012). Rowan seed production is very low, more comparable with fruit production amongst heavy-seeded tree species, like beech or oak (Francis 1983, Mosandl & Abt 2016, Gavranović et al. 2018) than silver birch or goat willow. This study revealed that a rowan fruit contains an average of 1.5-3.0 seeds (see also Sperens 1997a, Maier 2010). This means that a rowan seed tree can produce 2-67,626 seeds per year. The successful maintenance of rowan in forest ecosystems in spite of the low seed numbers may owe to directional seed dispersal by birds (Clark et al. 1998, Żywiec et al. 2013) (see chapter 7.1, p. 162). It would appear, therefore, that in the case of tree species with seed dispersed via zoochory the same high seed numbers produced by wind-dispersed trees are not required to achieve the same regeneration success. The reason for the high variation in seed production between mast and non-mast years maybe the result of resource depletion after fruit production (Gurnell 1993), as is the case in heavy-seeded tree species. Due to the generally low level of seed production, and the high annual variations of seed crops, rowan corresponds more closely to the heavy-seeded trees than to wind-dispersed pioneer tree species.

Pioneer trees do not generally exhibit high levels of seed production annually as is often assumed. There are large annual variations in seed production, similar to heavy-seeded tree species (Fischer et al. 2016). The variations are influenced by weather conditions. Low tree vitality, small dimensions and high competition pressure can also lead to low levels of seed production in pioneer tree species (Sarvas 1948, Maier 2010). Good seed years (mast years) occur in birch every 2-3 years on average (Sarvas 1948, Zerbe 2001) and every 2-5 years in rowan (Sperens 1997b, Zerbe 2001, Satake et al. 2004). The available findings for goat willow indicate that this may one species that does achieve high levels of seed production annually (see Ryvarden 1971, Brouwer & Stählin 1975, Lautenschlager 1994, Barsoum 2002, Gage & Cooper 2005, Kuzovkina & Quigley 2005, Argus 2006, Seiwa et al. 2008). This was also observed in this study. It is not only the frequency of masts, but also the seed quantities produced that differ between the three pioneer tree species. A goat willow seed tree produces 5-10 times more seeds than birches in mast years. In non-mast years the difference may be even greater, because goat willow is not as susceptible to the prevailing weather conditions as birch. The seed crops produced by rowan are always significantly lower and more strongly influenced by weather conditions than those of the wind-dispersed pioneer tree species.

This begs the question of how pioneer tree species compensate low levels of seed production in non-mast years, for example, through soil seed banks or seedling banks, to the extent that the general impression of high annual seed production and regeneration success could become so widely established.

Due to the differences between the pioneer tree species, the questions regarding the effectiveness of the ecological processes occurring within the individual regeneration cycles are discussed separately for the tree species in the following sections.

7.1.2 Ecological processes within the regeneration cycle of *Salix caprea*

Goat willow is not able to build up a soil seed bank, which confirms the third hypothesis (p. 11) underlying this study. This was shown by the results of the seed burial experiment, where no goat willow seeds were viable after 6 months storage in soil, although the initial germination capacity was 100 %. Numerous other studies found high germination capacities of 85-100 % for willow seeds and recorded short viability periods of 7-21 days after seed maturation and dispersal, because goat willow seeds have no food reserve (Juntilla 1976, Densmore & Zasada 1983, Lautenschlager 1994, Douglas 1995, Thompson et al. 1997, 1998, Skvortsov

1999, Bekker et al. 2000, Maroder et al. 2000, Barsoum 2002, Karrenberg & Suter 2003, Gage & Cooper 2005, Argus 2006, Schütt et al. 2011). To establish, goat willow seed must germinate and develop rapidly.

Investigations of the seedling banks of pioneer trees exceeded the scope of this study, but Juntilla (1976), Niiyama (1990), Lautenschlager (1994), Mihók et al. (2005), Schütt et al. (2011) and Richardson et al. (2014) mentioned the very low shade tolerance of willow seedlings and saplings, which require an irradiation level of more than 40 % of open land conditions. Prevailing light conditions in a closed forest are mostly between 1-20 % of above canopy light (Gralla et al. 1997, Wagner & Müller-Using 1997, Mihók et al. 2005, Huth 2009), which is not sufficient for willow seedlings to survive. Goat willow seedlings cannot grow or form a seedling bank under canopy cover and wait for favorable conditions to emerge; for example, after a windthrow event (Gralla et al. 1997, Wagner & Müller-Using 1997, Stancioiu & O'Hara 2006).

The annual regeneration success of goat willow depends solely on the quantity of the annual seed rain. This seed rain, therefore, would appear to represent a vital component of silvicultural measures targeting forest regeneration with goat willow given the following: (i) the comparatively minor reduction of seed numbers in non-mast years (seed crops are similarly high every year) compared to silver birch and rowan; (ii) the very high germination capacities; (iii) the long seed dispersal distances of more than 800 m, as described in chapter 2.4.2, p. 37 ff.; and (iv) the source-independent omnipresence of at least 20-45 seeds per m² (= no directionality) – which refutes hypothesis 2 (p. 10). Disturbed areas far from the nearest goat willows, or close to trees with low seed production, might still be regenerated by seed flow from goat willow at a considerable remove (Kikuchi et al. 2011, Trybush et al. 2012, Perdereau et al. 2014).

7.1.3 Ecological processes within the regeneration cycle of *Betula pendula*

The annual variability in silver birch seed production is significantly greater than in goat willow. In non-mast years the initial germination capacity of birch seeds is in most cases also significantly lower (Sarvas 1952, Bjorkbom 1967, Marquise 1969, Holm 1994). The initial percentage of birch seed germination has been found to range between 10-93 % (Sarvas 1952, Black & Wareing 1954, Holm 1994, Huth 2009). Given the limited and relief-induced mean dispersal distances of 80-380 m at high elevations and along the ridges in the Thuringian Forest (hypothesis 2a and 2b, p. 10), the findings of this study indicate that here non-mast years cannot be compensated by seed flow from other silver birch populations. The data from this study revealed no directionality to birch seed dispersal, and longer seed dispersal distances in the main wind direction. Holm (1994) found that groups of silver birch in proximity to one another usually exhibited synchronous behavior in seed production. Often there may not be close silver birch populations, however, as pioneer tree species are rarely found in managed conifer forests (Keidel et al. 2008, Heurich 2009, Brang et al. 2015). Therefore, silver birch seed rain cannot guarantee the successful natural regeneration of disturbed sites in non-mast years (Marquise 1969, Holm 1994).

Silver birch is able to build a short-term persistent soil seed bank (up to a maximum of 12 years), as assumed in the third hypothesis (p. 11, see also chapter 6.4.1, p. 141 ff. - Skoglund &Verwijst 1989, Thompson et al. 1997). Short-term persistent means that the soil seed bank needs a continuous seed rain input every few years. Given that good seed years occur every 2-3 years on average (Sarvas 1948, Zerbe 2001), silver birch seeds do not need to be able to persist in the soil for longer periods. The necessary replenishment of the silver birch soil seed bank with fresh seeds at an interval of at least every 5 years is well matched by the frequency of mast years in the species, as the results of this study (see chapter 3.5.1, p. 77 ff. and chapter 6.4.1, p. 141 ff.) and the findings presented by Granström (1987) and Skoglund & Verwijst (1989) showed. If seed trees are available but fructification fails in a particular year, the soil seed bank can still ensure successful regeneration.

Silver birch, like goat willow, cannot establish a seedling bank under the canopy due to its requirement for light exceeding 40 % of open-area irradiation (Marquise 1969, Gilbert et al. 2001, Portsmuth & Niinemets 2006, Huth 2009, 2015). Kobe et al. (1995) found that young yellow birch trees died within 2.5 years under light conditions below 9 %.

Although in a non-mast year silver birch can regenerate by virtue of the seed reserve in soil, germination of these buried seeds will not happen if there is no change to the prevailing environmental conditions (= disturbance). The seeds will not germinate in undisturbed soil (Granström 1987, Perala & Alm 1990). Under the right conditions, however, and contrary to the conclusions drawn by Hill (1979) and Heinrichs (2010), the regeneration of silver birch on disturbed sites depends not only on annual seed rain, but is also possible from the soil seed bank.

7.1.4 Ecological processes within the regeneration cycle of *Sorbus aucuparia*

Rowan trees must rely on a number of different strategies for successful regeneration. Rowan reacts much more sensitively to unfavorable weather conditions than silver birch and goat willow, resulting in an extreme reduction and sometimes failure of fruit production (Sperens

1997b, Satake et al. 2004, Żywiec et al. 2012). The 2-5 year intervals between good seed years (Sperens 1997b, Zerbe 2001, Satake et al. 2004) are also longer, and the seed dispersal distances significantly shorter, than for silver birch because of the reliance on endozoochorous dispersal. The reduced distances birds usually fly after eating limit effective seed dispersal to 30-100 m, which confirmed hypothesis 2a (p. 10, see also chapter 4.5, p. 95 ff. - Bakker et al. 1996, Jordano & Schupp 2000, Stiebel 2003, Holeksa & Żywiec 2005, Żywiec et al. 2013, Żywiec 2014).

Surprisingly, the proven ability of rowan to form a soil seed bank (see Raspé et al. 2000) was limited to a maximum of 3-4.5 years (hypothesis 3, p.11, see also chapter 6.4.1, p. 141 ff.). Rowan seeds without and with pulp stored in the soil for an average of 1-1.5 years, during which time embryo and seed coat dormancy are broken. After overcoming dormancy, the seeds usually germinate in the soil even without the occurrence of a disturbance (Granström 1987). As a consequence, the seed reserve in the soil needs constant replenishment, unlike silver birch. The presence of a sufficient accumulation of rowan seeds in the soil to serve as a seed reserve for the successful regeneration of disturbed sites cannot be assumed. In actual fact the 'seed bank' established by rowan seeds does not meet the definition of a soil seed bank (= seed reserve) (Fenner1985, Thompson et al. 1997, Berger et al. 2004, Bossuyt & Honnay 2008, Leck et al. 2008). Both the soil seed bank and seed rain are, therefore, not a reliable basis for the successful regeneration of disturbed sites with rowan.

Rowan can, however, build up a seedling bank under canopy cover (Holeksa & Żywiec 2005, Żywiec & Holeksa 2012). Rowan seedlings and saplings can survive under shelter at between 20-30 % of open-area irradiation levels (Prien 1964, Bartsch & Röhrig 2016). These seedlings can respond to the creation of better light conditions (e.g., through thinning or disturbances) with better growth up to a duration of 30 years (Gockel 2016). This ability to establish a seedling bank might be explained as an evolutionary adaptation to the dispersal vector 'frugivorous birds' and their behavior. Birds prefer protected, shaded forest edges and closed forests for defecation and avoid bright open areas lacking structural elements (McDonnell & Stiles 1983, Stimm & Böswald 1994, Gregor & Seidling 1997, Jordano & Schupp 2000, Stiebel 2003, Żywiec & Ledwoń 2008, Żywiec 2014). If the seedlings and saplings needed high light conditions to grow, the young trees would die under the shelter of the canopy and birds would not contribute to the successful regeneration of rowan. The germination of rowan seeds in undisturbed soil also makes sense in the context of its ability to create a seedling bank.

In summary, rowan regeneration depends on seed dispersal by birds, its ability to form a short-term soil seed bank but, above all, on its ability to establish a seedling bank under shelter, similar to the shade-tolerant climax species beech and silver fir (Szwagrzyk et al. 2001, Szymura 2005, Stancioiu & O'Hara 2006).

The results of the study showed that goat willow is the only one of the three pioneer tree species studied that meets the general assumptions made about the characteristics of pioneer tree species. The regeneration success of **goat willow** depends solely on the high annual seed production and long seed dispersal distances. Goat willow possesses no other regeneration abilities, such as the ability to establish a soil seed bank or a seedling bank.

Silver birch deviates from the general assumptions made about pioneer tree species and exhibits the characteristics of a hybrid. The high regeneration success of birch derives not only from the seed rain, which varies considerably between years and is dispersed shorter distances than goat willow seeds. This regeneration success stems also from a short-term persistent soil seed bank.

The properties exhibited by **rowan** during the regeneration cycle correspond more to shade-tolerant tree species than to the general assumptions made about light-demanding pioneer tree species. Compared to silver birch and goat willow, the rowan seed crop is significantly lower and the dispersal distances much shorter. Although rowan is able to build up a short-term soil seed bank, its strength as a species to quickly regenerate a disturbed site lies in its great ability to form a seedling bank. In general, however, the ecology of rowan (short-lived, low resistance to competition pressure, higher light requirement with increasing age, low site requirements during regeneration) corresponds to the characteristics of pioneer tree species (Raspé et al. 2000, Zerbe 2001, Gockel 2016).

All relevant aspects and stages of the regeneration cycles of the pioneer tree species are summarized in Appendix 4, p. iv.

In summary, the pioneer tree species studied have each adopted different regeneration mechanisms during evolution depending on the seed morphology and the dispersal vector characteristics. Every aspect within the regeneration cycle of each pioneer tree species is precisely coordinated, which has served to create the impression that their successful regeneration is the result of an abundant annual seed rain. In reality this applies only for goat willow. The perfect cooperation between all aspects of the regeneration mechanisms has ensured the successful regeneration and survival of the pioneer tree species studied up to now in spite of their removal from the forests in the past as part of forest management activities.

7.2. Conclusions for silviculture and management recommendations

The results of the study showed the great importance of the integration and tending of pioneer seed trees in managed forests, if natural regeneration is to be applied effectively as a management approach for disturbed sites. Since the storms Vivian and Wiebke in 1990, Lothar in 1999 and Kyrill in 2007, each of which resulted in extensive damage to forests in Germany, the exploitation of trees of pioneer species as pioneer forest or as filling-in material has become more widely practiced (Leder 2003, Leder et al. 2007, ThüringenForst 2013). To ensure the continued survival and establishment of seed-producing pioneer tree species in Germany's forests (in sufficient numbers and at appropriate spatial distributions) appropriate measures must be incorporate into silviculture. Silvicultural recommendations on the integration of pioneer species in stands through thinning and other measures are indispensable.

Management concepts targeting the natural regeneration of disturbed sites first necessitate knowledge of the seed tree numbers needed per hectare. To determine this number, the dispersal distances of seeds and the seed numbers deposited on the ground are crucial. It was shown in chapters 3.6, p. 74 ff. and chapter 2.6, p. 47 ff. that a minimum of 4-16 seed trees are needed per hectare in the case of silver birch and rowan, and 1-2 goat willow trees. The omnipresence of goat willow seeds makes a 'spatial optimisation' of seed trees unnecessary for the species. Although it would in theory be sufficient to have only one female and one male goat willow individual within an area 3-8 ha, 1-2 seed trees per hectare are recommended to account for any unforeseen losses to seed trees. Silver birch and rowan seed trees do require spatial optimization. Distances of 30-60 m between individuals are recommended to guarantee a homogeneous distribution of seeds.

However, the mere presence of a sufficient number of pioneer seed trees in managed forests does not guarantee successful seed dispersal to disturbed sites. Fructification and the successful production of large seed crops are necessary to ensure adequate seed rain for the regeneration of disturbed sites. On open areas, rowan and goat willow plants from seeding begin reproduction at 4-5 years (Zerbe 2001, Maier 2010, Roloff & Pietzarka 2010) and birch at 5-10 years (Atkinson 1992, Zerbe 2001). In forest areas birch reproduces for first time after 20-30 years (Zerbe 2001), rowan after 8-20 years (Prien 1964, Raspé et al. 2000, Zerbe 2001) and goat willow from the tenth year (Schirmer 2006). The contrasting onsets of fructification in forests and on open sites are indicative of the low competitiveness of pioneer tree species.

The strict allometric relationships between growth parameters and the individual seed crop of a tree described here are already established (Grisez 1975, Sato & Hiura 1998, Greene et al. 2004, Wagner et al. 2004, Huth 2009, Gockel 2016, DaPonte Canova 2018). Growth subject to high competition pressure and low light results in poor crown development, small dimensions (Hynynen et al. 2010), little to no fructification (Gockel 2016) and, in the worst cases, to the death of pioneer trees (Prien 1964, Maier 2010). For example, at high elevations and along ridges, birches with small crowns and low stability have to endure snow damage and often die after 30-50 years (Cameron 1996, Hering et al. 1999, Hynynen et al. 2010, Noack oral communication in 2015). Under such conditions, goat willow may already disappear from stands at the pole wood stage (Neumann 1981). Therefore, the control of space and interspecific competition to favor stem and crown dimensioning in pioneer seed trees is important.

Żywiec et al. (2012) observed that very favorable conditions for individual rowan seed trees can lead to local, spatially delineated and unsynchronized mast years. This means that silvicultural measures to limit competition may lead to higher seed production even in non-mast years (Huth 2009, Gockel 2016, Tiebel et al. 2019).

Pioneer tree species are short-lived tree species and do not usually exceed lifespans of 60-150 years (Atkinson 1992, Raspé et al. 2000, Schütt 2006). With the onset of senescence – age 40-60 years for birch, 30-50 years for goat willow – the aforementioned allometric relationship loses its tight correlation (Atkinson 1992, Schirmer 2006, Huth 2009, Hynynen et al. 2010). Therefore, individual pioneer trees are only a temporary presence in stands (Hering et al. 1999) and continuous promotion becomes necessary to ensure the recruitment of sufficient new flowering individuals.

Silvicultural measures to recruit new light-demanding seed trees of pioneer species involve the creation of gaps in otherwise closed (conifer) stands. The active promotion of existing seed trees is also recommended and measures to protect against browsing must be considered. Young pioneer trees are very often browsed and the bark of older individuals stripped (Prien 1964, Chantal & Granström 2007, Keidel et al. 2008, Gockel 2016). Around established seed trees regular regulation of the available space and pre-commercial thinning (= competitor removal) is necessary to increase crown development and individual tree stability. Pre-commercial thinnings should start in the thicket stage at the latest, independent of the pioneer tree species in question. Measures for the promotion of these trees should be implemented consistently until maturation (Prien 1964, Cameron 1996, Huth 2009, Gockel 2016). The earlier thinning occurs, the more pioneer tree species can respond (Cameron 1996, Schütt 2006, Gockel 2016). Rowan is the only one of these species that may still be able to respond to a first thinning measure at the age of 30 years (Gockel 2016), given its juvenile shade-tolerance. Prien (1964) found, however, that the demand for light already increases in rowan individuals of 8-10 years (= 3-4 m in height). Therefore, even for rowan, the first thinning should not happen too late. It should also be noted that for older trees a sudden release may also have negative effects where the trees can no longer adapt to the new conditions and die (Prien 1964, Cameron 1996).

Measures should generally only be carried out to promote dominant seed trees, to ensure that these selected individuals can persist in the sub-dominant or dominant stand layer (Prien 1964, Huth 2009). Where light-demanding pioneer seed tree species form the canopy alongside the target crop tree species, Hynynen et al. (2010) recommended space control and thinning at short intervals to maintain a crown ratio of 50 %, an indicator of vitality. To ensure good and vigorous crown development, thinning intervals for dominant birch trees should be carried out at intervals of 5-7 years (Cameron 1996). Prien (1964) and Gockel (2016) concluded that rowan requires relatively intense promotion with measures for space control to be implemented at intervals of 3-5 years. To correspond with cultivation intervals in regular forestry, a 5 year control interval is suggested for rowan, with the removal of stronger competitors recommended wherever necessary and possible. There are no existing thinning guidelines for goat willow growing in managed temporal forests. As a high light-demanding and fast-growing species with a high space requirement (Lautenschlager 1994, Argus 2006, Schirmer 2006, Schütt 2006, Gockel 2016), a control interval of 5 years is also suggested.

To ensure the efficiency and lower the intensity of silvicultural measures, it is recommended that treatment concepts be developed for groups of seed trees rather than for individual admixed seed trees in stands (Prien 1964, Cameron 1996, Hillebrand 1998, Huth 2009, Hynynen et al. 2010). Groups of seed trees reduce the pressure of interspecific competition and the number of thinning required. The adoption of seed tree groups also allows for larger distances between groups than is the case for individual seed trees. The seed shadows cast by seed trees in groups overlap, resulting in higher seed densities per square meter. Seed tree groups also guarantee a minimization of the risk of the loss of seed sources through mortality, for example, as a result of snow damage, insect calamity or senescence. The maintenance of sufficient numbers of fructifying seed trees of pioneer tree species in managed forests ensures that there is potentially sufficient seed rain to cater for disturbed sites.

The subsequent successful germination of seeds and establishment of seedlings required to ensure the reforestation of disturbed sites is tied to the availability of a sufficient number of microsites, optimal weather and soil conditions, and tolerable browsing pressure and herb competition (Marquise 1969, Junttila 1976, Densmore & Zasada 1983, Sacchi & Price 1992, Guthörl 1994, Cameron 1996, Young & Clements 2003, Schütt 2006, Huth 2009).

The scope for the natural regeneration of windthrown sites by pioneer tree species also has limits. The extent of seed dispersal, whether silver birch seeds dispersed on the wind or rowan dispersed by birds, is specific to the species. Endozoochorous dispersal of seeds further depends on the availability of structural elements on open sites (see chapter 4.4, p. 92 ff.). It is not possible to guarantee comprehensive natural regeneration of all pioneer tree species solely by seed rain across the whole extent of disturbed sites greater than 4 ha in size (see chapter 3.6, p. 74 ff. and chapter 4.5, p. 95 ff.). The following should, therefore, be observed: (i) pioneer seedlings already established on disturbed areas must be maintained; (ii) seed trees that survive disturbance events must be left on site; (iii) silver birch and rowan soil seed banks and (iv) rowan seedling banks should be exploited, if possible. On very large open areas of several hundred square meters or more, (v) natural regeneration must be supported by additional planting or seeding measures.

7.3 References

Abé S (2002): Seedling/sapling banks and their responses to forest disturbance. In: Nakashizuka T, Matsumoto Y (Eds): Diversity and interaction in a temperate forest community. Ecological Studies (Analysis and Synthesis) 158: 143-154.

Argus GW (2006): Guide to *Salix* (willow) in the Canadian Maritime Provinces (New Brunswick, Nova Scotia, and Prince Edward Island). E-Publishing, Ontario. URL: http://accs.uaa.alaska.edu/files/botany/publications/2006/GuideSalixCanadianAtlanticMaritime.pdf. Accessed 10 February 2018.

Atkinson MD (1992): *Betula pendula* Roth (*B. verrucosa* Ehrh.) and *B. pubescens* Ehrh. Journal of Ecology 80(4): 837-870.

Baier R, Meyer J, Göttlein A (2007): Regeneration niches of Norway spruce (*Picea abies* [L.] Karst.) saplings in small canopy gaps in mixed mountain forests of the Bavarian Limestone Alps. European Journal of Forest Research 126: 11-22.

Bakker JP, Poschlod P, Strykstra RJ, Bekker RM, Thompson K (1996): Seed banks and seed dispersal: important topics in restoration ecology. Acta Botanica Neerlandica 45(4): 461-490.

Barsoum N (2002): Relative contributions of sexual and asexual regeneration strategies in *Populus nigra* and *Salix alba* during the first years of establishment on a braided gravel bed river. Evolutionary Ecology 15: 255-279.

Bartsch N, Röhrig E (2016): Waldökologie - Einführung für Mitteleuropa. Springer-Verlag, Berlin, Heidelberg, 417 S.

Bastide JP, Vredenburch CLH van (1970): The influence of weather conditions on the seed production of some forest trees in the Netherlands. Acta Botanica Neerlandica 45: 461-490.

Baumann K (2007): Die Waldkiefer (*Pinus sylvestris*) – Baum des Jahres 2007. S. 140-144.

Bekker RM, Verwej GL, Bakker JP, Fresco LFM (2000): Soil seed bank dynamics in hayfield succession. Journal of Ecology 88: 594-607.

Berger TW, Sun B, Glatzel G (2004): Soil seed banks of pure spruce (*Picea abies*) and adjacent mixed species stands. Plant and Soil 264: 53-67

Bjorkbom JC (1967). Seedbed-preparation methods for paper birch. Northeastern Forest Experiment Station, Forest Service Research Paper NE-79. U.S. Department of Agriculture, Upper Darby, USA, 15 p.

Black M, Wareing PF (1954): Photoperiodic control of germination in seed of birch *(Betula pubescens* Ehrh.). Nature 174: 705-706.

Bossuyt B, Honnay O (2008): Can the seed bank be used for ecological restoration? An overview of seed bank characteristics in European communities. Journal of Vegetation Science 19: 875-884.

Bramlett DL (1990): *Pinus serotina* Michx. In: Burns RM, Honkala BH (Eds): Silvics of North America. Southeastern Forest Experiment Station, Asheville, NC, Forest Service. U.S. Department of Agriculture, pp. 470-475.

Brang P, Hilfiker S, Wasem U, Schwyzer A, Wolgemuth T (2015): Langzeitforschung auf Sturmflächen zeigt Potenzial und Grenzen der Naturverjüngung. Schweizerische Zeitschrift für Forstwesen 166(3): 147-158.

Brokaw NVL (1986): Seed dispersal, gap colonization, and the case of *Cecropia insignis*. In: Estrada A, Fleming TH (Eds): Frugivores and seed dispersal. Tasks for vegetation science 15. Springer, Dordrecht, pp. 323-331.

Brokaw NVL (1987): Gap-phase regeneration of three pioneer tree species in a tropical forest. Journal of Ecology 75: 9-19.

Brouwer W, Stählin A (1975): Handbuch der Samenkunde für Landwirtschaft, Gartenbau und Forstwirtschaft. DLG-Verlags-GmbH, Frankfurt (Main), 655 S.

Bucher HU (2008): *Abies alba* Miller, 1786. In: Schütt P, Weisgerber H, Schuck HJ, Lang U, Stimm B, Roloff A (Eds): Lexikon der Nadelbäume. WILEY-VCH Verlag GmbH & Co. KGaA, Weinheim, S. 1-18.

Burschel P, Huss J (1997): Grundriss des Waldbaus - Ein Leitfaden für Studium und Praxis. Pareys Studientexte 49. Blackwell Wissenschafts-Verlag, Berlin-Wien. 2. Aufl., 487 S.

Burse K-D, Schramm H-J, Geiling S, Meinhardt H, Schölch M (1997): Die forstlichen Wuchsbezirke Thüringens - Kurzbeschreibung, Mitteilungen der Landesanstalt für Wald und Forstwirtschaft Heft 13, 199 S.

Bushart M, Suck R (2008): Potenzielle natürliche Vegetation Thüringens, Schriftenreihe der Thüringer Landesanstalt für Umwelt und Geologie Nr. 78, 139 S.

Cameron AD (1996): Managing birch woodlands for the production of quality timber. Forestry 69: 357-371.

Chantal M, Granström A (2007): Aggregations of dead wood after wildfire act as browsing refugia for seedlings of *Populus tremula* and *Salix caprea*. Forest Ecology of Management 250: 3-8.

Chmelar J, Meusel W (1986): Die Weiden Europas. Ziemsen Verlag, Wittenberg. 3. Aufl., 144 S.

Clark JS, Macklin E, Wood L (1998): Stages and spatial scales of recruitment limitation in Southern Appalachian Forests. Ecological Monographs 68: 213-235.

Dalling JW, Swaine MD, Garwood NC (1998): Dispersal patterns and seed bank dynamics of pioneer trees in moist tropical forest. Ecology 79(2): 564-578.

Daniels J (2001): Ausbreitung der Moorbirke (*Betula pubescens* Ehrh. agg.) in gestörten Hochmooren der Diepholzer Moorniederung. Osnabrücker Naturwissenschaftliche Mitteilungen 27: 39-49.

DaPonte Canova G (2018): Regeneration ecology of anemochorous tree species *Qualea grandiflora* Mart. and *Aspidosperma tomentosum* Mart. of the cerrado Aguara Ñu located in the Mbaracayú Nature Forest Reserve (MNFR), Paraguay. Technical University of Dresden, Faculty of Environmental Science, PhD book, 160 p.

Densmore R, Zasada J (1983): Seed dispersal and dormancy patterns in northern willows: ecological and evolutionary significance. Canadian Journal of Botany 61: 3207-3216.

Dölle M, Schmidt W (2009): The relationship between soil seed bank, above-ground vegetation and disturbance intensity on old-field successional permanent plots. Applied Vegetation Science 12: 415-428.

Douglas DA (1995): Seed germination, seedling demography, and growth of *Salix setchelliana* on glacial river gravel bars in Alaska. Canadian Journal of Botany 73: 673-679.

Elmqvist T, Ericson L, Danell K, Salomonson A (1988): Latitudinal sex ratio variation in willows, *Salix* spp., and gradients in vole herbivory. Oikos 51(3): 259-266.

Erlbeck R (1998): Die Vogelbeere (*Sorbus aucuparia*) - ein Porträt des Baumes des Jahres 1997. In: Schmidt O (Hrsg.): Beiträge zur Vogelbeere. Berichte aus der Bayrischen Landesanstalt für Wald und Forstwirtschaft (LWF), Nr. 17: 2-14.

Fenner M (1985): Seed ecology. Chapman and Hall Ltd, London and New York, 151 p.

Fischer H, Huth F, Hagemann U, Wagner S (2016): Developing restoration strategies for temperate forests using natural regeneration processes. In: Stanturf JA (Ed): Restoration of Boreal and Temperate Forests. CRC Press, Boca Raton, pp. 103-164.

Francis JK (1983): Acorn production and tree growth of nuttall oak in a green-tree reservoir. Southern Forest Experiment Station, Forest Service Research Note SO-209. U.S. Department of Agriculture, Stoneville, Mississippi, 3 p.

Fries G (1984): Den frösådda björkens invandring på hygget (Immigration of birch into clearfelled areas). Sveriges Skogsvårdsförbunds Tidskrift 82: 35-39.

Gage EA, Cooper DJ (2005): Patterns of willow seed dispersal, seed entrapment, and seedling establishment in a heavily browsed montane riparian ecosystem. Canadian Journal of Botany 83: 678-687.

Gauer J, Aldinger E (2005): Waldökologische Naturräume Deutschlands: Forstliche Wuchsgebiete und Wuchsbezirke - mit Karte 1:1.000.000, Mitteilungen des Vereins für Forstliche Standortskunde und Forstpflanzenzüchtung, Nr. 43, 324 S.

Gautier-Hion A, Duplantier J-M, Quris R, Feer F, Sourd C, Decoux J.-P, Dubost G, Emmons L, Erard C, Hecketsweiler P, Moungazi A, Roussilhon C, and Tliiollay J-M (1985): Fruit characters as a basis of fruit choice and seed dispersal in a tropical forest vertebrate community. Oecologia 65: 324-337.

Gavranović1 A, Bogdan S, Lanšćak M, Čehulić I, Ivanković M (2018): Seed yield and morphological variations of beechnuts in four European beech (*Fagus sylvatica* L.) populations in Croatia. SEEFOR 9(1): 17-27.

Gilbert IR, Jarvis PG, Smith H (2001): Proximity signal and shade avoidance differences between early and late successional trees. Nature 411: 792-795.

Gockel S (2016): Wachstumsreaktionen einzeln eingemischter Vogelbeeren (*Sorbus aucuparia* L.) in Fichtenjungbeständen nach Freistellung. Technische Universität Dresden, Fakultät der Umweltwissenschaften, Tharandt, 348 S.

Google (2019): Karte Tharandter Wald. URL: https://www.google.de/maps/place/Tharandter+Wald/, Abruf: 23.07.2019.

Gralla T, Müller-Using B. Unden T, Wagner S (1997): Über die Lichtbedürfnisse von Buchenvoranbauten in Fichtenbaumhölzern des Westharzes. Forstarchiv 68(1): 51-58.

Granström A (1987): Seed viability of fourteen species during five years of storage in a forest soil. Journal of Ecology 75: 321-331.

Greene DF, Johnson EA (1997): Secondary dispersal of tree seeds on snow. Journal of Ecology 85(3): 329-340.

Greene DF, Canham CD, Coates KD, Lepage PT (2004): An evaluation of alternative dispersal functions for trees. Journal of Ecology 92: 758-766.

Gregor T, Seidling W (1997): 50 Jahre Vegetationsentwicklung auf einer Schlagfläche im osthessischen Bergland. Forstwissenschaftliches Cebtralblatt 116: 218-231.

Grisez T (1975): Flowering and seed production in seven hardwood species. Northeastern Forest Experiment Station, Forest Service Research Paper NE-315, Upper Darby, USA, 8 p.

Gurnell J (1993): Tree seed production and food conditions for rodents in an oak wood in southern England. Forestry 66(3): 291-315.

Guthörl V (1994): Zusammenhänge zwischen der Populationsdichte des Rehwildes (*Capreolus capreolus*, Linné 1758) und dem Verbißdruck auf die Waldvegetation. Zeitschrift für Jagdwissenschaft 40: 122-136.

Harcombe PA, White BD, Glitzenstein JS (1982): Factors influencing distribution and first-year survivorship of a cohort of beech (*Fagus grandifolia* Ehrh.). Castanea 47(2): 148-157.

Harper JL (1977): Population biology of plants. Academic Press, New York, 892 p.

Hecker U (Hrsg.) (1998): Bäume und Sträucher. BLV Verlagsgesellschaft mbH, München. 2. Aufl., 478 S.

Heinrichs S (2010): Response of the understorey vegetation to select cutting and clear cutting in the initial phase of Norway spruce conservation. Georg-August-University Göttingen, Faculty of Mathematics and Natural Sciences, PhD book, 177 p.

Hering S, Eisenhauer R, Irrgang S (1999): Waldumbau auf Tieflands- und Mittelgebirgsstandorten in Sachsen. Sächsische Landesanstalt für Forsten (LAF) (Hrsg.), 67 S.

Heurich M (2009): Progress of forest regeneration after a large-scale *Ips typographus* outbreak in the subalpine *Picea abies* forests of the Bavarian Forest National Park. Silva Gabreta 15(1): 49-66.

Hill MO (1979): The development of a flora in even-aged plantations. In: "The ecology of even-aged forest plantations" (Ford ED, Malcolm DC, Atterson J (Eds)). Institute of Terrestrial Ecology, Cambridge, UK, p. 175-192.

Hill MO, Stevens PA (1981): The density of viable seed in soils of forest plantations in upland Britain. Journal of Ecology 69: 693-709.

Hillebrand K (1998): Vogelbeere (*Sorbus aucuparia* L.) im Westfälischen Bergland - Wachstum, Ökologie, Waldbau-. Landesanstalt für Ökologie, Bodenordnung und Forsten/Landesamt für Agrarordnung NRW (Hrsg.), LÖBF-Schriftenreihe Bd. 15, 184 S.

Hilton GM, Packham JR (2003): Variation in the masting of common beech (*Fagus sylvatica* L.) in northern Europe over two centuries (1800-2001). Forestry 76(3): 319-328.

Holeksa J, Żywiec M (2005): Spatial pattern of a pioneer tree seedling bank in old-growth European subalpine spruce forest. Ekológia (Bratislava) 24(3): 263-276.

Holm SO (1994): Reproductive variability and pollen limitation in three *Betula* taxa in northern Sweden. Ecography 17(1): 73-81.

Hopfensperger KN (2007): A review of similarity between seed bank and standing vegetation across ecosystems. Oikos 116(9): 1438-1448.

Houle G, Payette S (1990) Seed Dynamics of *Betula alleghaniensis* in a deciduous forest of north- eastern North America. Journal of Ecology 78: 677-690.

Hughes JW, Fahey TJ (1988): Seed dispersal and colonization in a disturbed northern hardwood forest. Bulletin of the Torrey Botanical Club 115: 89-99.

Huth F, Wagner S (2006): Gap structure and establishment of Silver birch regeneration (*Betula pendula* Roth.) in Norway spruce stands. Forest Ecology and Management 229: 314-324.

Huth F (2009): Untersuchungen zur Verjüngungsökologie der Sand-Birke (*Betula pendula* Roth). Technischen Universität Dresden, Fakultät Forst-, Geo- und Hydrowissenschaften, Tharandt, book, 383 S.

Huth F (2015): Integration von Pionierbaumarten in die reguläre Bestandesbehandlung - Möglichkeiten und Grenzen. Vortrag zur Weiterbildung von ThüringenForst zum Thema: Risiken und Unsicherheiten durch den Waldumbau mit Pionierbaumarten mindern. Forstamt Frauenwald, 06.10.2015.

Huth F, Wehnert A, Tiebel K, Wagner S (2017): Direct seeding of Silver fir (*Abies alba* Mill.) to convert Norway spruce (*Picea abies* L.) forests in Europe: A review. Forest Ecology and Management 403: 61-78.

Hynynen J, Niemistö P, Viherä-Aarnio A, Brunner A, Hein S, Velling P (2010): Silviculture of birch (*Betula pendula* Roth and *Betula pubescens* Ehrh.) in northern Europe. Forestry 83: 103-119.

Iszkuło G, Boratyńska A, Didukh Y, Romaschenko K, Pryazhko N (2005): Changes of population structure of *Taxus baccate* L. during 25 years in protected area (Carpathians, western Ukraine). Polish Journal of Ecology 53(1): 13-23.

Jordano P, Schupp EW (2000): Seed disperser effectiveness: The quantity component and patterns of seed rain for *Prunus mahaleb*. Ecological Monographs 70: 591-615.

Junttila O (1976): Seed germination and viability in five *Salix* species. Astarte 9: 19-24.

Kandemir G, Kaya Z (2009): Technical Guidelines for genetic conservation and use of oriental beech (*Fagus orientalis*). EUFORGEN. Bioversity International, Italy. 6 p.

Karlsson M (2001): Natural regeneration of broadleaved tree species in southern Sweden - Effects of silvicultural treatments and seed dispersal from surrounding stands. Swedish University of Agricultural Sciences, Southern Swedish Forest Research Centre, Alnarp, PhD thesis, 44 p.

Karrenberg S, Suter M (2003): Phenotypic trade-offs in the sexual reproduction of *Salicaceae* from flood plains. American Journal of Botany 90: 749-754.

Keidel S, Meyer P, Bartsch N (2008): Untersuchungen zur Regeneration eines naturnahen Fichtenwaldökosystems im Harz nach großflächiger Störung. Forstarchiv 79: 187-196.

Kikuchi S, Suzuki W, Sashimura N (2011): Gene flow in an endangered willow *Salix hukaoana* (*Salicaceae*) in natural and fragmented riparian landscapes. Conservation Genetics 12: 79-89.

Kobe RK, Pacala SW, Silander JA Jr, Canham CD (1995): Juvenile tree survivorship as a component of shade tolerance. Ecological Application 5(2): 517-532.

Kohlermann L (1950): Untersuchungen über die Windverbreitung der Früchte und Samen mitteleuropäischer Waldbäume. Forstwissenschaftliches Centralblatt 69: 606-624.

Kollmann J (2000): Dispersal of fleshy-fruited species: a matter of spatial scale? Perspectives in Plant Ecology, Evolution and Systematics 3: 29-51.

Kremer BP (2010): Bäume und Sträucher. Eugen Ulmer KG, Stuttgart (Hohemheim). 2. Aufl. 380 S.

Kullmann L (1993): Tree limit dynamics of *Betula pubescens* ssp. *tortuosa* in relation to climate variability: evidence from central Sweden. Journal of Vegetation Science 4: 765-772.

Kuzovkina YA, Quigley MF (2005): Willows beyond wetlands: uses of *Salix* L. species for environmental projects. Water, Air, & Soil Pollution 162: 183-204.

Lautenschlager D (1994): Die Weiden von Mittel- und Nordeuropa. Birkhäuser Verlag, Basel. überarbeitete Auflage. 171 S.

Leck MA, Parker VT, Simpson RL (2008): Seedling ecology and evolution. University Press, Cambridge, UK, 514 p.

Leder B (1992): Weichlaubhölzer- Verjüngungsökologie, Jugendwachstum und Bedeutung in Jungbeständen der Hauptbaumarten Buchen und Eiche. Schriftenreihe der Landesanstalt für Forstwissenschaft Nordrhein-Westfalen, Sonderband, 416 S.

Leder B (2003): „Natürliche Wiederbewaldung nach Fichtenwindwurf 1990". LÖBF-Mitteilungen 2: 40-43.

Leder B, Asche N, Dame G, Gertz M, Hein F, Kreienmeier U, Naendrup G, Sondermann P, Spelsberg G, Stemmer M, Wagner H-C, Wrede Ev (2007): Empfehlung für die Wiederbewaldung der Orkanflächen in Nordrhein-Westfalen, Arnsberg. Landesbetrieb Wald und Holz Nordrhein-Westfalen, Lehr- und Versuchsforstamt Arnsberger Wald (Hrsg), 71 S.

Lertzman KP, Krebs CJ (1991): Gap-phase structure of a subalpine old-growth forest. Canadian Journal of Forest Research 21: 1730-1741.

Lima RAF de, Prado PI, Martini AMZ, Fonseca LJ, Gandolfi S, Rodrigues RR (2012): Improving methods in gap ecology: revisiting size and shape distributions using a model selection approach. Journal of Vegetation Science 24(3): 484-495.

Liu Q, Hytteborn H (1991): Gap structure, disturbance and regeneration in a primeval Picea abies forest. Journal of Vegetation Science 2: 391-402.

Löf M (2000): Establishment and growth in seedlings of *Fagus sylvatica* and *Quercus robur*: influence of interference from herbaceous vegetation. Canadian Journal of Forest Research 30: 855-864.

MacCarthaigh D, Spethmann W [Hrsg.] (2000): Krüssmans Gehölzvermehrung. Parey-Verlag, Berlin. 435 S.

Maier J (2010): *Sorbus aucuparia* LINNÉ. In: Roloff A, Weisberger H, Lang U, Stimm B (Hrsg.): Bäume Mitteleuropas. WILEY-VCH Verlag GmbH & Co. KGaA, Weinheim, 1. Aufl., S. 223-238.

Maroder HL, Prego IA, Facciuto GR, Maldonado SB (2000): Storage behaviour of *Salix alba* and *Salix matsudana* seeds. Annals of Botany 86: 1017-1021.

Marquise DA (1969): Silvical requirements for natural birch regeneration. Northeastern Forest Experiment Station, Forest Service, Upper Darby, USA, pp 40-49.

Matlack GR (1989): Secondary dispersal of seed across snow in *Betula lenta*, a gap-colonizing tree species. Journal of Ecology 77(3): 853-869.

McDonnell MJ, Stiles EW (1983): The structural complexity of old field vegetation and the recruitment of bird-dispersed plant species. Oecologia 56: 109-116.

Mihók B, Gálhidy L, Kelemen K, Standovár T (2005): Study of gap-phase regeneration in a managed beech forest: Relations between tree regeneration and light, substrate features and cover of ground vegetation. Acta Silvatica et Lignaria Hungarica 1: 25-38.

Moles AL, Falster DS, Leishman MR, Westoby M (2004): Small-seeded species produce more seeds per square metre of canopy per year, but not per individual per lifetime. Jornal of Ecology 92: 384-396.

Mosandl R, Abt A (2016): Waldbauverfahren in Eichenwäldern gestern und heute. AFZ-Der Wald 20: 28-32.

Neumann A (1981): Die mitteleuropäischen *Salix*-Arten. Mitteilungen der forstlichen Bundes-Versuchsanstalt Wien, Heft 134, 152 S.

Niiyama K (1990): The role of seed dispersal and seedling traits in colonization and coexistence of *Salix* species in a seasonally flooded habitat. Ecological Research 5: 317-331.

Övergaard R, Gemmel P, Karlsson M (2007): Effects of weather conditions on mast year frequency in beech (*Fagus sylvatica* L.) in Sweden. Forestry 80(5): 555-565.

Perala DA, Alm AA (1990): Reproductive ecology of birch: A review. Forest Ecology and Management 32(1): 1-38.

Perdereau AC, Kelleher CT, Douglas GC, Hodkinson TR (2014): High levels of gene flow and genetic diversity in Irish populations of *Salix caprea* L. inferred from chloroplast and nuclear SSR markers. Plant Biology 14: 1-12.

Popadyuk RV, Smirnova OV, Evstigneev OI, Yanitskaya TO, Chumatchenko SI, Zaugolnova LB, Korotkov VN, Chistyakova AA, Khanina LG, Komarov AS (1995): Current state of broad-leaved forests in Russia, Belorussia, Ukraine: historical development, biodiversity, structure and dynamic. Preprint, PRC RAS, Pushchino, 73 p.

Portsmuth A, Niinemets Ü (2006): Interacting controls by light availability and nutrient supply on biomass allocation and growth of *Betula pendula* and *B. pubescens* seedlings. Forest Ecology and Management 227: 122-134.

Prien S (1964): Untersuchungen über waldbauliche und holzkundliche Eigenschaften der Eberesche (*Sorbus aucuparia* L.). book der TU Dresden, Sektion Forstwirtschaft Tharandt, 227 S.

Raspé O, Findlay C, Jacquemart A-L (2000): *Sorbus aucuparia* L. Journal of Ecology 88: 910-930.

Reif A, Gärtner S (2007): Die natürliche Verjüngung der laubabwerfenden Eichenarten Stieleiche (*Quercus robur* L.) und Traubeneiche (*Quercus petraea* Liebl.) - eine Literaturstudie mit besonderer Berücksichtigung der Waldweide. Waldoekologie online 5: 79-116.

Richardson J, Isebrands JG, Ball JB (2014): Ecology and physiology of poplars and willows. In: Isebrands JG, Richardson J (Eds): Poplars and willows: trees for society and the environment. FAO, pp. 92-123.

Röhrig E, Bartsch N, Lüpke Bv (2006): Waldbau auf ökologischer Grundlage. Eugen Ulmer KG, Stuttgart (Hohenheim). 7. Aufl., 479 S.

Roloff A, Pietzarka U (2010): *Betula pendula* ROTH. In: Roloff A, Weisberger H, Lang U, Stimm B (Hrsg.): Bäume Mitteleuropas. WILEY-VCH Verlag GmbH & Co. KGaA, Weinheim. 1. Aufl., S. 61-75.

Runkle JR (1992): Guidelines and sample protocol for sampling forest gaps. Pacific Northwest Research Station, Forest Service, General Technical Report PNW-GTA-283. US Department of Agriculture, 44 p.

Runkle JR, Yetter TC (1987): Treefalls revisited: gap dynamics in the southern Appalachians. Ecology 68(2): 417-424.

Ryvarden L (1971): Studies in seed dispersal. I. Trapping of Diaspores in the Alpine Zone at Finse, Norway. Norwegen Journal of Botany 18: 215-226.

Sacchi CF, Price PW (1992): The relative roles of abiotic and biotic factors in seedling demography of arroyo willow (*Salix lasiolepis: Salicaceae*). American Journal of Botany 79: 395-405.

Safford LO, Jacobs RD (1983): Paper birch. In: Burns RM (Eds): Silvicultural systems for the major forest types of the United States, U.S. Department of Agriculture Washington, D. C., Agriculture Handbook No. 445: 145-147.

Sarvas R (1948): A research on the regeneration of birch in South Finland. Communicationes Instituti Forestalis Fenniae 35(4): 82-91.

Sarvas R (1952): On the flowering of birch and the quality of seed crop. Communicationes Instituti Forestalis Fenniae 40: 1-35.

Satake A, Bjørnstad ON, Kobro S (2004): Masting and trophic cascades: interplay between rowan trees, apple fruit moth, and their parasitoid in Southern Norway. Oikos 104(3): 540-550.

Sato H, Hiura T (1998): Estimation of overlapping seed shadows in a northern mixed forest. Forest Ecology of Management 104: 69-76.

Schiechtl HM (1992): Weiden in der Praxis - Die Weiden Mitteleuropas, ihre Verwendung und ihre Bestimmung. Patzer-Verlag, Berlin-Hannover, 130 S.

Schirmer R (2006): *Salix alba* Linné. In: Schütt P, Weisgerber H, Schuck HJ, Lang U, Stimm B (Eds): Enzyklopädie der Laubbäume. Nikol Verlagsgesellschaft mbH & Co KG, Hamburg, 535-550.

Schütt P, Stimm B (2001): *Salix caprea* L., 1753. Enzyklopädie der Holzgewächse 26. Erg.Lfg. 12/01, 8 S.

Schütt P (2006): *Salix caprea* LINNÉ, 1753. In: Schütt P, Weisgerber H, Schuck HJ, Lang U, Stimm B, Roloff A (Eds): Enzyklopädie der Laubbäume. Nikol Verlagsgesellschaft mbH & Co. KG, Hamburg, S. 551-558.

Schütt P, Stimm B (2006): *Pinus sylvestris* L., 1753. Enzyklopädie der Holzgewächse 45. Erg. Lfg. 9/06, 33 S.

Schütt P (2008): *Taxus baccata* Linné, 1753. In: Schütt P, Weisgerber H, Schuck HJ, Lang U, Stimm B, Roloff A (Eds): Lexikon der Nadelbäume. WILEY-VCH Verlag GmbH & Co. KGaA, Weinheim, S. 573-583.

Schütt P, Schuck HJ, Stimm B (Hrsg.) (2011): Lexikon der Baum- und Straucharten - Das Standardwerk der Forstbotanik. Nikol Verlagsgesellschaft mbH & Co KG, Hamburg, 581 S.

Seiwa K, Tozawa M, Ueno N, Kimura M, Yamasaki M, Maruyama K (2008): Roles of cottony hairs in directed seed dispersal in riparian willows. Plant Ecology 198: 27-35.

Silveira FAO, Fernandes GW, Lemos-Filho JP (2013): Seed and seedling ecophysiology of neotropical *Melastomatacea*: implications for conservation and restoration of savannas and rain forest. Annals of the Missouri Botanical Garden 99(1): 82-99.

Smithsonian (undated): Cecropia insignis. Smithsonian Tropical Research Institute, URL: https://biogeodb.stri.si.edu/bioinformatics/croat/specie/Cecropia%20insignis,e,n. Abruf: 21.10.2019.

Skoglund J, Verwijst T (1989): Age structure of woody species populations in relation to seed rain, germination and establishment along the river Dalälven, Sweden. Vegetation 82: 25-34.

Skvortsov AK (1999): Willows of Russia and adjacent countries - Taxonomical and geographical revision, University of Joensuu, Faculty of mathematics and natural sciences report series, Finland, 307 p.

Sork VL (1993): Evolutionary ecology of mast-seeding in temperate and tropical oaks (*Quercus* spp.). Vegetatio 107/108: 133-147.

Sperens U (1997a). Long-term variation in, and effects of fertiliser addition on, flower, fruit and seed production in the tree Sorbus aucuparia (*Rosaceae*). Ecography 20: 521-534.

Sperens U (1997b): Fruit production in *Sorbus aucuparia* L. (*Rosaceae*) and pre-dispersal seed predation by the apple fruit moth (*Argyresthia conjugella* Zell.). Oecologia 110: 368-373.

Stancioiu PT, O'Hara KL (2006): Regeneration growth in different light environments of mixed species, multiaged, mountainous forests of Romania. European Journal of Forest Research 125: 151-162.

Stiebel H (2003): Frugivorie bei mitteleuropäischen Vögeln. Carl-von-Ossietzki-Universität Oldenburg, Fakultät für Mathematik und Naturwissenschaften, Wilhelmshaven, book, 219 S.

Stimm B, Böswald K (1994): Die Häher im Visier. Zur Ökologie und waldbaulichen Bedeutung der Samenausbreitung durch Vögel. Forstwissenschaftliches Centralblatt 113: 204-223.

Szwagrzyk J, Szewczyk J, Bodziarczyk J (2001): Dynamics of seedling banks in beech forest: results of a 10-year study on germination, growth and survival. Forest Ecology and Management 141: 237-250.

Szymura TH (2005): Silver fir sapling bank in seminatural stand: Individuals architecture and vitality. Forest Ecology and Management 212 (1-3): 101-108.

Thompson K, Bakker J, Bekker R (1997): The soil seed banks of North West Europe: methodology, density and longevity. Cambridge University Press, Cambridge, UK, 276 p.

Thompson K, Bakker JP, Bekker RM, Hodgson JG (1998): Ecological correlates of seed persistence in soil in the north-west European flora. Journal of Ecology 86: 163-169.

ThüringenForst (2013): Wiederbewaldung im Team. URL: http://www.thueringenforst.de/de/forst/aktuelles/wiederbewaldung/. Stand: 08.04.2013, Abruf: 19.12.2013.

ThüringenForst-AöR (2014): Forsteinrichtungsdaten.

Thüringer Klimaagentur (2015a): Witterungsbericht Frühjahr 2015. Thüringer Landesanstalt für Umwelt und Geologie (TLUG) - Thüringer Klimaagentur, 5 S.

Thüringer Klimaagentur (2015b): Witterungsbericht Sommer 2015. Thüringer Landesanstalt für Umwelt und Geologie (TLUG) - Thüringer Klimaagentur, 7 S.

Tiebel K, Wehnert A, Huth F, Erefur C, Bergstens U, Wagner S (2019): Fruktifikation der Salweide am Beispiel Nordschwedens. AFZ-Der Wald 12: 25-27.

Trybush SO, Jahodová Š, Čížková L, Karp A, Hanley SJ (2012): High levels of genetic diversity in *Salix viminalis* of the Czech Republic as revealed by microsatellite markers. Bioenergy Research 5: 969-977.

Volkert B, Reif A (2010): Naturverjüngung der Traubeneiche (*Quercus petraea* (Matt.) Liebl.) in Hecken im Zartener Becken (Schwarzwald). Mitteilungen des badischen Landesvereins für Naturkunde und Naturschutz 21:1-23.

Wagner S, Müller-Using B (1997): Ergebnisse der Buchen-Voranbauversuche im Harz unter besonderer Berücksichtigung der lichtökologischen Verhältnisse. Schriftenreihe, Landesanstalt für Ökologie, Bodenordnung und Forsten/Landesamt für Agrarordnung Nordrhein-Westfalen, Nr. 13: 17-30.

Wagner S, Wälder K, Ribbens E, Zeibig A (2004): Directionality in fruit dispersal models for anemochorous forest trees. Ecological Modelling 179: 487-498.

Wohlgemuth T, Nussbaumer A, Burkart A, Moritzi M, Wasem U, Moser B (2016): Muster und treibende Kräfte der Samenproduktion bei Waldbäumen. Schweizerische Zeitschrift für Forstwesen 167(6): 316-324.

Wright SJ, Trakhtenbrot A, Bohrer G, Detto M, Katul GG, Horvitz N, Muller-Landau HC, Jones FA, Nathan R (2008): Understanding strategies for seed dispersal by wind under contrasting atmospheric conditions. PNAS 105(49): 19084-19089.

Young JA, Clements CD (2003): Seed germination of willow species from a desert riparian ecosystem. Journal of Range Management 56: 496-500.

Zerbe S (2001): On the ecology of *Sorbus aucuparia* (Rosaceae) with special regard to germination, establishment and growth. Polish Botanical Journal 46(2): 229-239.

Żywiec M, Ledwoń M (2008): Spatial and temporal patterns of rowan (*Sorbus aucuparia* L.) regeneration in West Carpathian subalpine spruce forest. Plant Ecology 194: 283-291.

Żywiec M, Holeksa J, Ledwoń M (2012): Population and individual level of masting in a fleshy-fruited tree. Plant Ecology 213: 993-1002.

Żywiec M, Holeksa J, Wesołowska M, Szewczyk J, Zwijacz-Kozica T, Kapusta P, Bruun HH (2013): *Sorbus aucuparia* regeneration in a coarse-grained spruce forest - a landscape scale. Journal of Vegetation Science 24: 735-743.

Żywiec M (2014): Seedling survival under conspecific and heterospecific trees: the initial stages of regeneration of *Sorbus aucuparia*, a temperate fleshy-fruit pioneer tree. Annales Botanici Fennici 50: 361-371.

Table of appendix

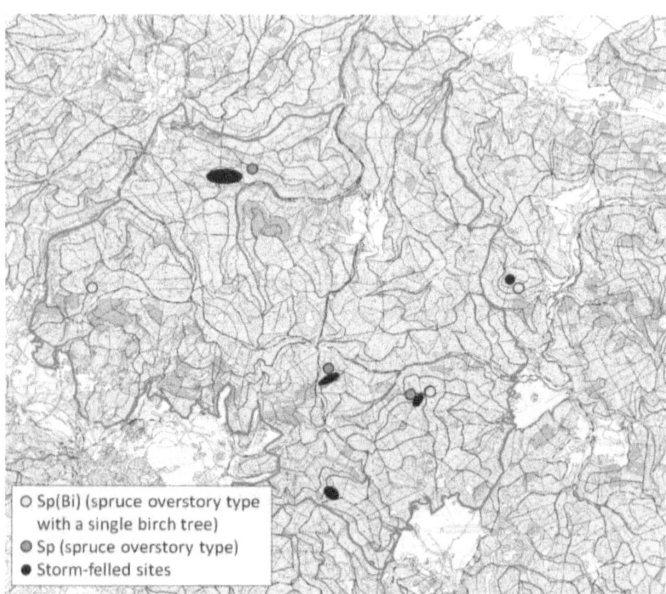

Appendix 1 Locations of the sites of the seed dispersal and soil seed bank investigations in the Thuringian Forest (Reference: modified from original forest map provided by ThüringenForst-AÖR 2014).

Appendix 2 Location of the sites of the soil seed bank investigation in the Tharandter Forest (Reference: Google 2019).

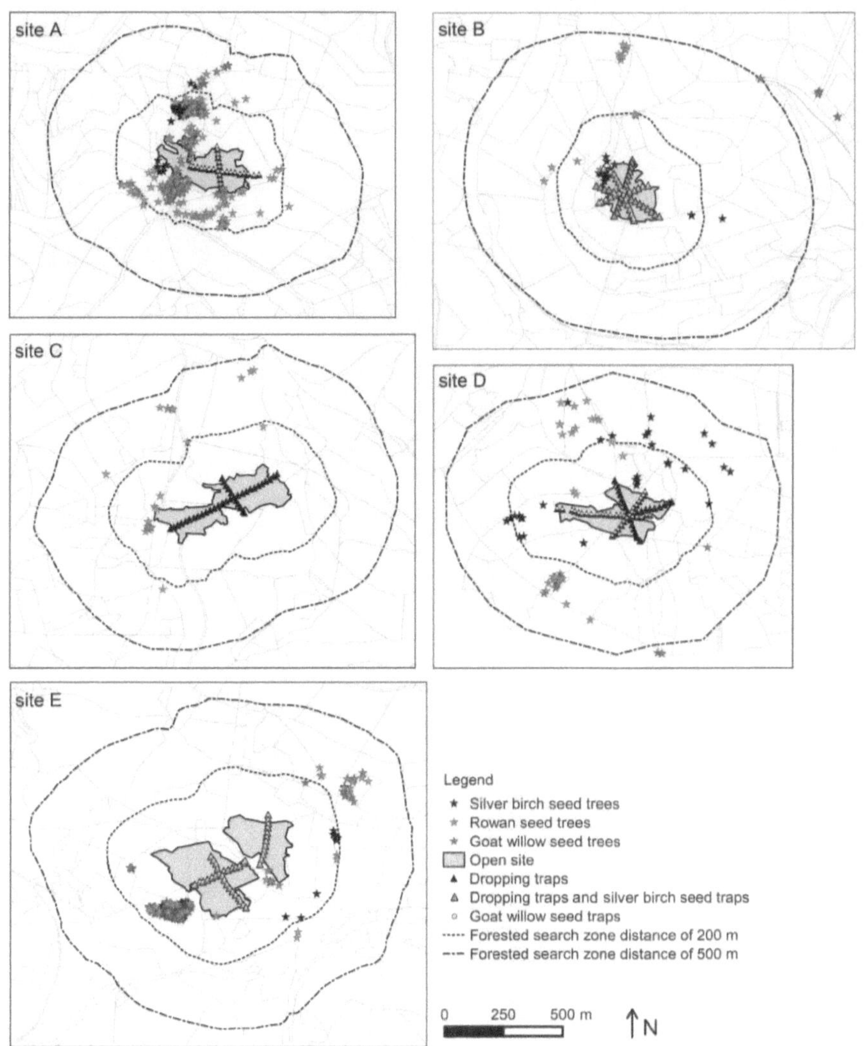

Appendix 3 Experimental study designs for the seed traps on the windthrown sites A-E in the Thuringian Forest with crossing line transects.

Appendix 4 Overview of the life histories and regeneration strategies (seed production, soil seed bank, seedling bank) of silver birch (*Betula pendula* Roth), goat willow (*Salix caprea* and rowan (*Sorbus aucuparia* L.) (References can be found in reference list in chapter 7 'General discussion').

	Silver birch	Goat willow	Rowan	References
Longevity (maximum) [years]	90–120 (180)	60	80–150	Atkinson 1992, Raspé et al. 2000, Zerbe 2001, Schütt 2006, Maier 2010, Roloff & Pietzarka 2010
Maximal tree height [m]	25–30	12–15	15–20	Atkinson 1992, Schiechtl 1992, Raspé et al. 2000, Zerbe 2001, Schütt 2006, Hynynen et al. 2010, Roloff & Pietzarka 2010
End of height growth of mature trees [years]	60	20–25	25–30	Prien 1964, Schütt 2006
Plant/flower	monoecious	dioecious	monoecious	Atkinson 1992, Raspé et al. 2000, Argus 2006, Schütt 2006
Pollination	wind	wind and insects	insects	Atkinson 1992, Raspé et al. 2000, Argus 2006
Vector of seed dispersal	anemochore	anemochore	zoochore	Atkinson 1992, Zerbe 2001, Argus 2006, Schütt 2006
Frequency of high seed production - 'mast years' [years]	2–3	unknown	2–5	Sarvas 1948, Sperens 1997b, Zerbe 2001, Satake et al. 2004
First flowering [years]				
Open sites	(2) 5–10	4–5	4–5	Atkinson 1992, Leder 1992, Raspé et al. 2000, Zerbe 2001, Schirmer 2006, Maier 2010, Roloff &Pietzarka 2010
Within forest	30	10	8–20	Prien 1964, Raspé et al. 2000, Zerbe 2001, Schirmer 2006, Maier 2010, Roloff & Pietzarka 2010
Seed production per tree [number]	-	1.2–22.3 million	-	Tiebel et al. 2019
Mast year	3.7–10 million	unknown	12,900–67,600	Perala &Alm 1990, Popadyuk et al. 1995, Sperens 1997b, Wagner et al. 2004, Huth 2009, Żywiec et al. 2012, Żywiec 2014
Non-mast year	40,000–1.5 million	unknown	3–7,900	this study, Sperens 1997b, Denisow 2007 (cited inHuth 2009), Żywiec et al. 2012, Żywiec 2014
Seed dispersal distance [m]				
Dispersal distances (maximum)	37–125 (192)	200–1,500 (3,000)	30–100 (550)	this study, Sarvas 1948, Hughes & Fahey 1988, Leder 1992, Bakker et al. 1996, Jordano &Schupp 2000, Daniels 2001, Karlsson 2001, Stiebel 2003, Wagner et al. 2004, Gage & Cooper 2005, Holeksa & Żywiec 2005, Schirmer 2006, Huth 2009, Żywiec et al. 2013, Żywiec 2014
Mean dispersal distance-uphill	(30) 87–96	*no differences*	unknown	this study, Hill et al. 1981
Mean dispersal distance-downhill	367–380		unknown	this study

Areas of deposited high seed numbers to seed tree [m]	25–50	up to 25	up to 40	this study, Sarvas 1948, Fries 1984, Skoglund & Verwijst 1989, Ryvarden 1971, Cameron 1996, Gage & Cooper 2005, Żywiec & Ledwoń 2008, Żywiec et al. 2013
Possibility of secondary seed dispersal	yes	yes	no	Matlack 1989, Greene & Johnson 1997, Gage & Cooper 2005
Influence of seed dispersal distance by following factors				
Azimuth direction (for entire dispersal period)	yes (no) - by wind	yes (no) - by wind	yes - by preference of birds	this study, Wagner et al. 2004, Gage & Cooper 2005, Wright et al. 2008, Huth 2009
Relief inclination	yes	no	unknown	this study
Position of seed tree (slope, valley, open area, edge, forest)	yes	no	yes	this study, McDonnel & Stiles 1983, Jordano & Schupp 2000, Stiebel 2003
Distance between seed tree and disturbed area	yes	yes (< 50 m) no (> 50 m)	yes	this study, Jordano & Schupp 2000, Stiebel 2003, Holeksa & Żywiec 2005, Żywiec et al. 2013, Żywiec 2014
Structural elements on disturbed area	-	-	yes	this study, McDonell & Stiles 1983, Kollmann 2000
Minimal number of necessary seed trees (n/ha)	4–16	1–2 (+1 male individual)	4–16	this study, Sarvas 1948, Safford & Jacobs 1983
Soil seed bank	yes	no	yes	
type	short-term to long-term persistent	transient	transient-short term persistent	this study, Bakker et al. 1996, Thompson et al. 1997, Bekker et al. 2000, Raspé et al. 2000, Dölle & Schmidt 2009
storage capacity (maximum) [years]	2–7 (13)	1	1–2 (5)	this study, Sarvas 1952, Hill 1979, Granström 1987, Skoglund & Verwijst 1989, Leder 1992, Thompson et al. 1997, Erlbeck 1998
Seedling bank	no	no	yes	Holeksa & Żywiec 2005, Żywiec & Holeksa 2012
Maximum height growth of saplings [m/year]	0.6–1	0.7–1	0.6–0.8	Prien 1964, Leder 1992, Hecker 1998, Zerbe 2001, Schütt 2006, Chantal & Granström 2007
Light requirement of seedlings (minimum) [open-area radiation level]	(> 30 %) > 40 %	(> 10 %) > 40 %	(< 20 %) > 20–30 %	Prien 1964, Marquise 1969, Juntilla 1976, Niiyama 1990, Kobe et al. 1995, Gilbert et al. 2001, Mihók et al. 2005, Portsmuth & Niinemets 2006, Richardson et al. 2014, Bartsch & Röhrig 2016
Thinning intervals of treatments for optimal growth	5–7	5	5 (3)	this study, Prien 1964, Cameron 1996, Gockel 2016

Declaration

Erklärung zur Eröffnung des Promotionsverfahrens

1. Hiermit versichere ich, dass ich die vorliegende Arbeit ohne unzulässige Hilfe Dritter und ohne Benutzung anderer als der angegebenen Hilfsmittel angefertigt habe; die aus fremden Quellen direkt oder indirekt übernommenen Gedanken sind als solche kenntlich gemacht.

2. Bei der Auswahl und Auswertung des Materials sowie bei der Herstellung des Manuskripts habe ich Unterstützungsleistungen von folgenden Personen erhalten:

An der konzeptionellen Entwicklung des übergeordneten Themas, den einzelnen Themenschwerpunkten (Kapitel 2-6) und des Versuchsdesigns waren Sven Wagner, Franka Huth und Nico Frischbier beteiligt.

Nico Frischbier von ThüringenForst und Sonja Gockel, als Koordinatorin im Rahmen des Modellprojektes "Waldumbau in den mittleren, Hoch- und Kammlagen des Thüringer Waldes", waren wichtige Kooperationspartner vor Ort in Thüringen. Sie waren sehr hilfreich bei der Auswahl der Untersuchungsflächen, der Koordination von Maßnahmen (mit den Waldarbeitern und Förstern) und der Bereitstellung von Ressourcen, wie Luftbildern, Kartenwerk etc.

Die Etablierung des Versuchsdesigns (Einmessung der Transekte) wurde mit Hilfe durch Antje Karge, Franka Huth, Alexandra Wehnert und Angelika Otto durchgeführt.

Anna Viktoria August, Klaus Tiebel und Kathrin Tiebel haben bei der Herstellung der Weidensamenfallen, Kotfallen und den Säckchen für die Bodensamenbankproben im Eingrabungsexperiment geholfen. Weiterhin halfen Antje Karge, Franka Huth, Alexandra Wehnert, Jörg Wollmerstädt, Julia Möhring und Anna Victoria August bei den turnusmäßigen Leerungen der Samenfallen. Anna Victoria August war zu diesem Zweck als Studentische Hilfskraft angestellt. Die Gewinnung der Bodenproben zur Untersuchung der Bodensamenbank im Thüringer Wald wurden durch Antje Karge, Franka Huth, Angelika Otto, Jörg Wollmerstädt und Jan Wilkens unterstützt. Im Rahmen einer Graduierungsarbeit (Bachelorarbeit) übernahm Julian Heinrich, unter meiner Anweisung, die Bodenprobengewinnung im Tharandter Wald und Datenerhebung im Gewächshaus. Die genetische Untersuchung zur Nachkommenschaftsanalyse von Salweide führten Ludger Leinemann und Bernhard Hosius von ISOGEN durch. Robert Schlicht und Sven Wagner waren maßgeblich an der Methodenentwicklung zur geostatistischen Auswertung der räumlichen Ausbreitung der Salweidensamen beteiligt.

Die Strukturierung der Manuskripte und deren Inhalte wurden mit den jeweiligen, artikelbezogenen Koautoren (Sven Wagner, Franka Huth, Alexandra Wehnert, Antje Karge, Nico Frischbier, Ludger Leinemann, Bernhard Hosius, Robert Schlicht, Julian Heinrich) diskutiert (siehe Kapitel 2-6). Des Weiteren haben alle Koautoren den ihrerseits betreffenden Artikel Korrektur gelesen.

3. Weitere Personen waren an der geistigen Herstellung der vorliegenden Arbeit nicht beteiligt. Insbesondere habe ich nicht die Hilfe eines kommerziellen Promotionsberaters in An-
spruch genommen. Dritte haben von mir weder unmittelbar noch mittelbar geldwerte Leistun-
gen für Arbeiten erhalten, die im Zusammenhang mit dem Inhalt der vorgelegten book stehen.

4. Die Arbeit wurde bisher weder im Inland noch im Ausland in gleicher oder ähnlicher Form
einer anderen Prüfungsbehörde vorgelegt und ist – sofern es sich nicht um eine kumulative
book handelt – auch noch nicht veröffentlicht worden.

5. Sofern es sich um eine kumulative book gemäß § 10 Abs. 2 der Promotionsordnung der
Fakultät Umweltwissenschaften handelt, versichere ich die Einhaltung der dort genannten
Bedingungen.

6. Ich bestätige, dass ich die Promotionsordnung der Fakultät Umweltwissenschaften der
Technischen Universität Dresden anerkenne.

Tharandt, 29.11.2019

Katharina Tiebel

Publications of the book

Manuscripts of book

TIEBEL, K.; HUTH, F.; WAGNER, S. (submitted): Do birch and rowan establish soil seed banks sufficient to compensate for a lack of seed rain after forest disturbance? Plant and Soil

TIEBEL, K.; HUTH, F.; WEHNERT, A.; WAGNER, S. (under Review): The impact of structural elements on storm-felled sites on endozoochorous seed dispersal by birds - a case study. iForest

TIEBEL, K.; HUTH, F.; FRISCHBIER, N.; WAGNER, S. (pre print): Restrictions on natural regeneration of storm-felled spruce sites by silver birch (*Betula pendula* Roth) through limitations in fructification and seed dispersal. European Journal of Forest Research 2020: https://doi.org/10.1007/s10342-020-01281-9

TIEBEL, K.; LEINEMANN, L.; HOSIUS, B.; SCHLICHT, R.; FRISCHBIER, N.; WAGNER, S. (2019): Seed dispersal capacity of *Salix caprea* L. assessed by seed trapping and parentage analysis. European Journal of Forest Research: 138(3): 495-511. https://doi.org/10.1007/s10342-019-01186-2

TIEBEL, K.; HUTH, F.; WAGNER, S. (2018): Soil seed banks of pioneer tree species in European temperate forests: a review. iForest 11: 48-57. 10.3832/ifor2400-011

Further publications of the study topic

TIEBEL, K.; HUTH, F.; WAGNER, S. (2019): Movement of pioneer tree species from forest into storm-felled areas by seed dispersal. Science meets practice. Book of Abstracts - 49[th]Annual Meeting of the Ecological society of Germany, Austria and Switzerland 2019, Gesellschaft für Ökologie, Münster, pp. 462.

TIEBEL, K.; WEHNERT, A.; HUTH, F.; EREFUR, C.; BERGSTENS, U.; WAGNER, S. (2019): Fruktifikation der Salweide am Beispiel Nordschwedens. AFZ-Der Wald 12: 25-27.

TIEBEL, K.; LEINEMANN, L.; HOSIUS, B.; FRISCHBIER, N.; WAGNER, S. (2018): Natural restoration of disturbed forests by *Salix caprea* L. In: Muraoka, Y.; Tschirf, R. (Ed.): Ecology – Meeting the scientific challenges of a complex world. Book of Abstracts - 48th Annual Meeting of the Ecological Society of Germany, Austria and Switzerland. Verhandlungen der Gesellschaft für Ökologie Band 47/48, Berlin/Vienna, pp. 157. ISSN 0171-1113 https://www.gfoe-conference.de/download/Book_of_Abstracts.pdf

TIEBEL, K.; KARGE, A.; HUTH, F.; WEHNERT, A.; WAGNER, S. (2017): Naturverjüngungspotenziale von Pionierbaumarten für die Wiederbewaldung von Sturmwurfflächen nutzen. Forstarchiv 88 (3): 138. Beitrag zur Tagung der der Sektion Waldbau im Deutschen Verband Forstlicher Forschungsanstalten (DVFFA), Bad Soden-Salmünster

TIEBEL, K.; KARGE, A.; HUTH, F.; WEHNERT, A.; WAGNER, S. (2017): Strukturelemente fördern die Samenausbreitung durch Vögel. AFZ-Der Wald 20:24-27.

TIEBEL, K.; HUTH, F.; WAGNER, S. (2015): How to catch different types of dispersed pioneer tree seeds in open areas successfully. Ecology for a sustainable future. Book of Abstracts - 45[th]Annual Conference 2015, Gesellschaft für Ökologie, Göttingen, pp. 237 - 238.

Danksagung

Die vorliegende Arbeit wurde von der Deutschen Bundestiftung Umwelt (DBU) finanziert, der ich an dieser Stelle für das gewährte Stipendium und der damit verbundenen Möglichkeit zur Promotion recht herzlich danken möchte. Ebenso danke ich der Graduiertenakademie für ihre finanzielle Unterstützung während der Abschlussphase meiner book und zum Forschungsaufenthalt in Schweden. Die finanzielle Unterstützung von ThüringenForst - namentlich durch Dr. Nico Frischbier - ermöglichte mir eine intensivere Beschäftigung mit der Pionierbaumart Salweide. Besonders möchte ich auch Herrn Wolfgang Arenhövel von ThüringenForst für die unverhoffte Möglichkeit zur finanziellen Durchführung der genetischen Untersuchungen an Salweide danken.

Mein besonderer Dank gebührt meinem betreuenden Hochschullehrer, Herrn Prof. Sven Wagner. Erst durch seine Anregung zu diesem Thema und schlussendlich dessen Überlassung war mir eine Promotion im Fach Waldbau möglich. Es kam vor, dass ich manchmal den Wald vor lauter Bäumen nicht sah und Probleme unlösbar erschienen. Ich möchte mich daher an dieser Stelle recht herzlich für die zahlreichen fachlichen Diskussionen und Anregungen, für die Hilfsbereitschaft zur Überwindung von statistischen Hürden und die Beseitigung des einen oder anderen Problems bedanken. All das hat zum Gelingen der Arbeit beigetragen.

Herrn Prof. Ammer und Frau Prof. Berger danke ich für die Bereitschaft zur Übernahme der weiteren Gutachten.

Ganz besonders danke ich Dr. Franka Huth, die mir als „Neuling" erklärt hat, wie das Leben als „Wissenschaftler" tatsächlich funktioniert und man sich durch den universitären Dschungel kämpft. Sie war es auch, die mich immerzu ermuntert hat, meiner Forschungsneugier und Interessen nachzugehen und über den Tellerrand der abgesteckten Methodik meines Promotionsthemas hinaus zu schauen und neue Ideen mutig umzusetzen. Ohne ihren wertvollen Zuspruch und Unterstützung wäre die vorliegende Arbeit deutlich kürzer ausgefallen.

ThüringenForst, insbesondere Dr. Sonja Gockel und Dr. Nico Frischbier, bin ich für die Flächenbereitstellung und -stilllegung für die Dauer meiner Untersuchungen zu großem Dank verpflichtet. Sie übernahmen dankenswerter Weise ebenfalls die Kommunikation, Koordination und Absprachen mit den Revierförstern. Nico Frischbier war ebenfalls stets für mich da, wenn ich Fragen zu den Untersuchungsflächen hatte, Wünsche bezüglich neuer Forschungsideen äußerte oder Unterstützung von Seiten ThüringenForst brauchte. Auch die fachlichen Diskussionen, seine Hilfestellungen und wohlgemeinten Ratschläge waren sehr wertvoll für mich.

Für die tatkräftige Unterstützung bei den Aufnahmen im Thüringer Wald danke ich ganz besonders Antje Karge. Diese frohen Stunden waren sehr wertvoll für mich, da sie eine willkommene Abwechslung zu den zahlreichen Wochen allein im Wald darstellten. An dieser Stelle möchte ich ebenfalls Angelika Otto, Jörg Wollmerstädt, Fanka Huth, Alexandra Wehnert, Anna Viktoria Augst, Julia Möhring, Jan Wilkens und den Waldarbeitern von ThüringenForst für ihre Hilfe bei den Probennahmen, sowie den Auf- und Abbauten des Versuchsdesigns bedanken.

Alexandra Wehnert danke ich ganz persönlich für ihren absolut unermüdlichen Einsatz und ihre grenzenlose Arbeitsbereitschaft während der Untersuchungen in Schweden. Ohne ihre Hilfe wäre der Aufenthalt bei weitem nicht so angenehm und reibungslos verlaufen. Und ich bezweifle, dass ich ohne ihre Hilfe einer meiner Forschungsideen hätte umsetzen können.

Ebenso möchte ich mich bei meinen Mitstreitern und Freunden an der Professur für Waldbau Giovanna Da Ponte, Manuela Böhme, Anastasia Wallraf, Jan Wilkens und Man Hung Bui für den wertvollen fachlichen Gedankenaustausch und die aufmunternden Gespräche bedanken.

Dr. David Bulter Manning und Dr. Ulrike Hagemann danke ich für ihre unermüdliche Arbeit mit der Verbesserung meiner englischsprachigen Manuskripte und deren wertvolle Tipps.

Ein Dankeschön für die hilfreichen Hinweise zu statistischen Fragen geht an Dr. Robert Schlicht.

Ohne Rückhalt und die tatkräftige Unterstützung meiner Eltern wäre diese Arbeit kaum möglich gewesen. Vermutlich wäre ich heute noch mit der Herstellung der Kotfallen beschäftigt und zu Fuß unterwegs, um meine Flächen zu erreichen. Sie haben mich immerwährend unterstützt, all meine Höhen und Tiefen, die mit der Arbeit einhergingen, ertragen und mir durch stressige Phasen mit ihrem Zuspruch geholfen.

Ariane van Aken war es, die akribisch über orthografische Fehler dieser Arbeit gewacht hat und mich in der letzten Phase meiner Promotionszeit immer wieder aufgebaut und motiviert hat. Ich danke ihr dafür.